SpringerBriefs in Applied Sciences and Technology

Computational Intelligence

Series Editor

Janusz Kacprzyk, Systems Research Institute, Polish Academy of Sciences, Warsaw, Poland

SpringerBriefs in Computational Intelligence are a series of slim high-quality publications encompassing the entire spectrum of Computational Intelligence. Featuring compact volumes of 50 to 125 pages (approximately 20,000-45,000 words), Briefs are shorter than a conventional book but longer than a journal article. Thus Briefs serve as timely, concise tools for students, researchers, and professionals.

Pedro Alberto Oliveira Paiva
José Pedro Ponte Mota da Costa
Filipe Parrado de Azevedo
Ricardo Miguel Ferreira Martins

Analog Integrated Circuit Design Under PVT Conditions

Efficient Reinforcement and Transfer Learning Techniques

 Springer

Pedro Alberto Oliveira Paiva
Instituto de Telecomunicacões/Instituto
Superior Técnico
University of Lisbon
Lisbon, Portugal

José Pedro Ponte Mota da Costa
Instituto de Telecomunicacões/Instituto
Superior Técnico
University of Lisbon
Lisbon, Portugal

Filipe Parrado de Azevedo
Instituto de Telecomunicacões/Instituto
Superior Técnico
University of Lisbon
Lisbon, Portugal

Ricardo Miguel Ferreira Martins
Instituto de Telecomunicacões/Instituto
Superior Técnico
University of Lisbon
Lisbon, Portugal

ISSN 2191-530X   ISSN 2191-5318 (electronic)
SpringerBriefs in Applied Sciences and Technology
ISSN 2625-3704   ISSN 2625-3712 (electronic)
SpringerBriefs in Computational Intelligence
ISBN 978-3-032-19408-4   ISBN 978-3-032-19409-1 (eBook)
https://doi.org/10.1007/978-3-032-19409-1

Preface

The evolution of the electronics industry and the increasing demand for consumer products have impacted on the design of integrated circuits (ICs). Deep nanometer technologies allowed an increased complexity of ICs, providing smaller and more power-efficient solutions, but consequently became more challenging to design. The IC market has seen substantial growth, valued at 681.05 billion USD in 2024, and is projected to grow to 2062.59 billion USD in 2032. This growth underscores the importance of addressing design challenges in the IC industry, which impacts various sectors like education, health care, and transportation. The development of these systems is crucial to meet society's evolving needs, driven by advancements in global communications and technology integration. When it comes to designing these systems, digital circuits are usually chosen to implement most of the functionalities due to the robust nature of handling digital signals, easier to reuse, and having many automated design tools within its design flow. However, most systems rely on interaction with the real world, and that is where analog, radio-frequency and millimeter wavelength (mmWave) ICs come into play. Systems where these two types of circuits are employed are denominated mixed-signal systems-on-a-chip, and can be found everywhere.

In the past years, automatic simulation-based sizing approaches became essential in designing these analog, radio-frequency and mmWave IC blocks for modern applications to ensure their robustness. However, optimizations considering process, voltage, and temperature (PVT) corners or layout still pose unprecedented challenges in applying these tools due to the high simulation times and different simulator convergence issues. Therefore, the work presented in this book addresses the automatic sizing of analog ICs assisted by machine and deep learning. The first part of this work presents a comparative study of five different approaches for integrating PVT conditions, along with several reward functions, within state-of-the-art reinforcement learning-driven sizing methodologies. Results reveal the trade-offs of each approach, and the best-performing one for this problem is highlighted and discussed to facilitate more research activities within this field. Furthermore, considering these PVT corners during traditional simulation-based techniques is also incurring prohibitive optimization times. Therefore, in the second part of this work, an additional technique is explored, where deep learning is used to assist a simulation-based sizing tool, via independent and parallel PVT performance regressors that bypass the simulator. These performance regressors are shallow artificial neural networks, where transfer learning from nominal to corner conditions was employed to avoid the need to acquire an expensive and time-consuming PVT training dataset, and an online Bayesian-assisted incremental learning is used to refine each model with accurate simulator data. This methodology is vastly tested on the design of millimeter wavelength ICs. When compared with full simulation-based synthesis, it reduced the workload of the circuit simulator up to 69% while achieving a speed-up factor of $3.4 \times$ and requiring less 86% dataset generation effort when compared with most recent PVT regressors.

Finally, the authors would like to express gratitude for the financial support that made this work possible. The work developed in this book was supported by national funds through FCT—Fundação para a Ciência e a Tecnologia, I.P., and, when eligible, co-funded by EU funds under project/support UID/50008/2025—Instituto de Telecomunicações, with DOI identifier 10.54499/UID/50008/2025, and also, by project ACTON, with DOI identifier 10.54499/2023.11981.PEX.

This book is organized into four chapters.

Chapter 1 presents an introduction to the analog, radio-frequency and mmWave IC design area and discusses how the advances in machine and deep learning can enhance existing electronic design automation tools in the analog spectrum.

Chapter 2 presents a comprehensive study of the available tools for analog design automation. Focusing on most recent works where machine learning techniques are applied to analog IC sizing.

Chapter 3 studies the incorporation of PVT corner conditions into reinforcement learning-based analog IC sizing, as it remains a challenging task. Therefore, a comparative study of five different approaches for integrating PVT conditions is conducted, where each approach begins with the same conditions to ensure a fair comparison, and is evaluated within the same circuit topology, comparing agent steps, number of simulations, execution time and final sizing functional behavior. Results reveal the trade-offs of each, and the best-performing one for this problem is highlighted and discussed to facilitate more research activities within this field.

Chapter 4 considers PVT corners during traditional simulation-based synthesis of mmWave ICs in nanometer technologies, which is incurring prohibitive optimization times. An innovative research towards the automation of mmWave IC design by using deep learning to assist a simulation-based sizing tool, via independent and parallel PVT performance regressors that bypass the simulator, is presented. There, transfer learning from nominal to corner conditions avoids the need for expensive PVT data, and an online Bayesian-assisted incremental learning is used to refine each model with accurate simulator data.

Lisbon, Portugal

Pedro Alberto Oliveira Paiva
José Pedro Ponte Mota da Costa
Filipe Parrado de Azevedo
Ricardo Miguel Ferreira Martins

Competing Interests The authors have no competing interests to declare that are relevant to the content of this manuscript.

Contents

1 Introduction 1
  1.1 Electronic Design Automation 1
  1.2 Analog, RF and mmWave IC Design Flow 2
  1.3 Machine Learning and Analog IC Sizing 3
  1.4 Conclusions 4
  References 4

2 State-of-the-Art 7
  2.1 Knowledge-Based Sizing 7
  2.2 Optimization-Based Sizing 7
    2.2.1 Equation-Based Evaluation 8
    2.2.2 Simulation-Based Evaluation 8
  2.3 Machine Learning in Analog/RF IC Sizing Automation 10
    2.3.1 Speeding-Up Simulation-Based Sizing with Artificial
         Neural Networks 10
    2.3.2 Reinforcement Learning in Analog/RF IC Sizing 12
    2.3.3 Transfer Learning in Analog/RF IC Design 19
  2.4 Conclusions 20
  References 20

**3 PVT Corner Conditions Within Reinforcement Learning-Based
Analog IC Sizing** 23
  3.1 Contributions 23
  3.2 Environment Overview 24
    3.2.1 Action Space 24
    3.2.2 Observation Space 25
    3.2.3 Reward Functions 26
  3.3 PVT-Inclusive Approaches 28
    3.3.1 Brute Force Approach 28
    3.3.2 Progressive Overload Approach 28

3.3.3   Reduced Corner Set Approach 30

3.3.4   Scheduler Approach 30

3.3.5   Worst-Case Approach 31

3.4 Agent Overview 31

3.4.1   Agent Algorithm 32

3.4.2   Implementation Details 33

3.5 Experimental Results 33

3.5.1   Case Studies 34

3.5.2   Reward Functions 36

3.5.3   PVT Corner-Inclusive Approaches 40

3.6 Conclusions and Future Research Directions 58

3.6.1   Conclusions 58

3.6.2   Future Work 59

References 59

4 PVT Corner Analog IC Sizing Optimizations Boosted
by ANN-Based Performance Regressors and Transfer Learning 61

4.1 Contributions 61

4.2 Proposed Methodology: TL-Enhanced PVT Regressors 62

4.2.1   TL-Enhanced PVTR Framework 62

4.2.2   Case Studies: 28 GHz Low-Noise Amplifiers 64

4.2.3   Design Specifications and PVT Corners 64

4.2.4   Transferability Analysis Between TT and PVT Domains 66

4.3 Development of the TT Performance Regressors 68

4.3.1   General In/Out Structure 68

4.3.2   Datasets 68

4.3.3   Hyperparameter Tuning 69

4.3.4   Evaluation and Final Model Configurations 72

4.4 Optimization Results 73

4.4.1   TL-Enhanced PVTR Framework as a Full Simulator
Surrogate 74

4.4.2   Controlled TL-Enhanced PVTR Framework 77

4.5 Conclusions and Future Research Directions 88

4.5.1   Conclusions 88

4.5.2   Future Work 89

References 90

Chapter 1
Introduction

1.1 Electronic Design Automation

As stated, IC industry is experiencing an unprecedented increase in demand for electronic devices, extending beyond consumer electronics into sectors such as healthcare, automotive and security, with designers developing increasingly more complex, power-efficient ICs. These systems commonly integrate analog and digital sections, where most components are combined into a single chip. Though several of the functionalities are implemented using digital signal processing circuitry, the analog sections provide the bridge between the digital components and physical world, and their design suffers from disproportional difficulties compared to digital. Furthermore, the stringent time-to-market constraints and rising development costs make electronic systems design highly challenging, thereby underscoring the paramount importance of developing electronic design automation (EDA) tools to accelerate and automate the design process. A vast number of computer-aided design (CAD) and EDA tools have been developed to cope with new capabilities offered by the integration technologies, nonetheless, the discrepancy between the analog and digital IC design tools is still significant. Unlike digital design, which can be translated into the form of Boolean expressions, the design of analog, RF and mmWave sections is less systematic, relying more on knowledge-based techniques and designer intervention, despite usually occupying only a small fraction of these SoCs. These difficulties have motivated the development of novel methodologies to automate and speed-up circuit design on the analog spectrum, with significant advances being made over the last few years in both academia and industry.

P. A. Oliveira Paiva et al., *Analog Integrated Circuit Design Under PVT Conditions*,
SpringerBriefs in Applied Sciences and Technology,
https://doi.org/10.1007/978-3-032-19409-1_1

1.2 Analog, RF and mmWave IC Design Flow

Recent years have seen significant advancements in large scale IC design due to automation mechanisms, however, design in the analog spectrum still primarily follows the flow systematized by Gielen and Rutenbar [2], which is presented in Fig. 1.1. This design flow involves a sequence of top-down steps, repeated from the system to the device level, and bottom-up steps, associated with layout generation and verification. The top-down flow focuses on optimizing system performance by progressively refining higher-level abstract representations into detailed, implementable designs. This strategy aims to enhance first-time success rates and decrease overall design time. However, real-world factors, such as layout parasitics and process variations often require multiple iterations to achieve a viable design.

The complexity of the system dictates the number of hierarchies in the design flow and, even though there is no globally accepted representation of the design architecture, it usually can be divided into two main paths: (1) top-down electric synthesis path—which includes the topology selection and specification translation, followed by a design verification; and (2) bottom-up physical synthesis path—which involves layout generation and respective layout extraction, finishing with a design verification.

Moreover, the high pressure in today's market for ultralow-power consumptions, larger communication rates and broader bandwidths have intensified the critical role of RF and mmWave IC design in deep nanometric integration technologies for both IoT and 5G telecommunication broadbands. Their design is not only inherently difficult due to the high complexity and demanding performances required by these systems, but also by the: impact of layout parasitics at gigahertz frequencies; dependence on unreliable models of passive devices; and by the fact that integration in deep

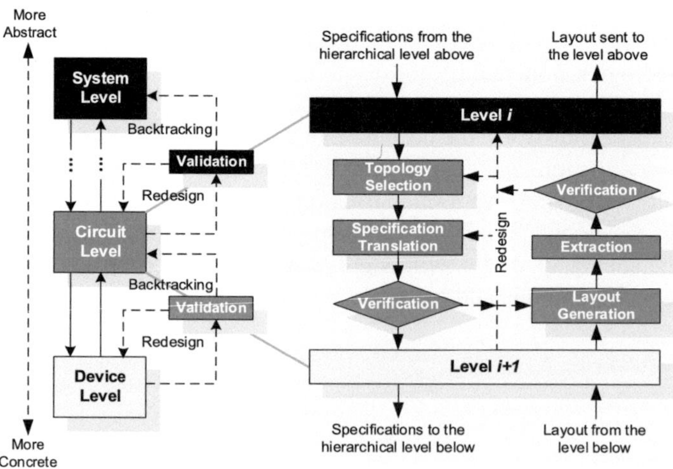

Fig. 1.1 Hierarchical levels and design tasks of analog design flow (Adapted from [2])

nanometer technologies triggers variability issues and non-idealities not present in older technology nodes.

Achieving first-pass fabrication success while avoiding costly redesign cycles and minimizing post-fabrication tuning has become a primary goal in RF and mmWave IC design. Established CAD tools provide environments that support this flow, although largely through manual effort. However, the traditional trial-and-error approach is no longer feasible due to the increasing complexity of interactions, which seldom lead to optimal RF and mmWave designs. Despite recent efforts in the field, automatic analog IC design frameworks have not yet reached the push button stage, with most modern techniques incurring in high computational resources and extensive design cycles. Overcoming these challenges requires ongoing improvement and development of new EDA tools, which remains an active area of research.

1.3   Machine Learning and Analog IC Sizing

Chapter 2 of this book will summarize the different approaches proposed over the years to automate the sizing task of analog ICs, from the traditional knowledge and equation-based approaches to the most sophisticated simulation-based optimizations [3]. The later has gained a prominent role in analog IC sizing automation as it exploits off-the-shelf circuit simulators, and performance evaluation via simulation is one of most well established concepts in the overall design flow. These automatic analog, RF and mmWave IC sizing methods aim to find the best set of device sizes by iterating over tentative guesses and evaluating their impact on circuit performance. This process is shown to produce usable designs, but it is still slow, and reuse usually involves new optimization runs. Moreover, in order to design robust IC blocks, process, voltage, and temperature (PVT) corners have to be mandatorily considered.

However, due to the time and computational cost of such circuit simulations, machine learning (ML) algorithms have been recently leveraged to help such techniques. ML is essentially how a computer improves its capabilities by analyzing past experiences. This area of artificial intelligence (AI) has been chosen to solve problems in computer vision, speech recognition, and natural language processing, among others because it can be easier to train a system showing examples of what the output should be given a particular input than to anticipate all possible responses for all inputs. Similarly, recent advances in ML and deep learning (DL) techniques have been offering new alternatives to the design automation of ICs not only on the mapping from devices' sizes to circuits' performances [4–6], i.e., replacing the simulator, but also on a broader set of design problems [7–11]. These include applications such as the inverse sizing problem, i.e., the mapping from specifications to the devices' sizes [12, 13], layout generation [14–17], fault testing [18], and so on. Therefore, other ML-based solutions for analog IC sizing beyond traditional simulation-based have also been proposed, and are here explored and analyzed, with a particular focus on reinforcement learning (RL) and transfer learning-based training. Furthermore, incorporating process, voltage, and temperature (PVT) corners into analog IC design is more critical than ever, as circuits are increasingly required to operate under the most adverse conditions. However, this topic has received limited attention in recent academic research, with only a handful of works reflecting its growing importance [19, 20].

1.4 Conclusions

While the demand for new analog ICs for ever-increasing challenging specifications keeps building momentum, their design automation is still lacking and needs to be addressed. In this book, we explore how ML/DL can be used to increase the effectiveness of analog, RF and mmWave automatic sizing. Chapter 3 presents a comparative study of five different approaches for integrating PVT conditions, along with several reward functions, within state-of-the-art RL-driven sizing methodologies. Results reveal the trade-offs of each approach, and the best performing one for this problem is highlighted and discussed to facilitate more research activities within this field. While Chap. 4 uses DL to assist a simulation-based sizing tool, via independent and parallel PVT performance regressors that bypass the simulator. These performance regressors are shallow artificial neural networks, where transfer learning from nominal to corner conditions was employed to avoid the need to acquire an expensive and time-consuming PVT training dataset, and an online Bayesian-assisted incremental learning is used to refine each model with accurate simulator data. This methodology is vastly tested on the design of millimeter wavelength ICs. When compared with full simulation-based synthesis, it reduced the workload of the circuit simulator up to 69% while achieving a speed-up factor of $3.4\times$, and requiring less 86% dataset generation effort when compared with most recent PVT regressors.

References

1. The Business Research Company (2025) Integrated Circuits Global Market Report 2025 Tech Rep. Available: https://www.thebusinessresearchcompany.com/report/integrated-circuits-global-market-report. Accessed 12 October 2025
2. Gielen GGE, Rutenbar RA (2000) Computer-aided design of analog and mixed-signal integrated circuits. Proc IEEE 88(12):1825–1854
3. Póvoa R et al (2014) LC-VCO automatic synthesis using multi-objective evolutionary techniques. In: IEEE international symposium on circuits and systems (ISCAS)
4. Wolfe G, Vemuri R (2003) Extraction and use of neural network models in automated synthesis of operational amplifiers. IEEE Trans Comput Aided Des Integr Circ Syst 22(2):198–212
5. Alpaydin G, Balkir S, Dundar G (2003) An evolutionary approach to automatic synthesis of high-performance analog integrated circuits. IEEE Trans Evol Comput 7(3):240–252
6. Liu H, Singhee A, Rutenbar R, Carley L (2002) Remembrance of circuits past: macromodeling by data mining in large analog design spaces. In: Proceedings 2002 design automation conference, pp 437–442
7. Mina R, Jabbour C, Sakr G (2022) A review of machine learning techniques in analog integrated circuit design automation. Electronics MDPI 11(3):435

8. Martins R, Lourenço N (2023) Analog integrated circuit routing techniques: an extensive review. IEEE Access 11:35965–35983

9. Wang C, Yang F, Zhu K (2024) AI-enabled layout automation for analog and RF IC: current status and future directions. In: Proceedings of the IEEE international symposium on radio-frequency integration technology

10. Maji S, Budak A, Poddar S, Pan D (2024) Toward end-to-end analog design automation with ML and data-driven approaches (Invited Paper). In: Proceedings of the 29th Asia and South Pacific design automation conference

11. Martins R (2025) A survey of machine and deep learning techniques in analog integrated circuit layout synthesis. Microelectron MDPI 1(1, 2)

12. Azevedo F, Lourenço N, Martins R (2025) Comprehensive application of denoising diffusion probabilistic models towards the automation of analog integrated circuit sizing. Expert Syst Appl 290:128414

13. Eid P, Azevedo F, Lourenço N, Martins R (2025) Using denoising diffusion probabilistic models to solve the inverse sizing problem of analog integrated circuits. AEU Int J Electron Commun 195:155767

14. Zhu K et al (2019) Genius route: a new analog routing paradigm using generative neural network guidance. In: Proceedings of the IEEE/ACM international conference on computer-aided design

15. Gusmão A, Horta N, Lourenço N, Martins R (2022) Scalable and order invariant analog integrated circuit placement with attention-based graph-to-sequence deep models. Expert Syst Appl 207:117954

16. Gusmão A, Póvoa R, Horta N, Lourenço N, Martins R (2022) DeepPlacer: a custom integrated OpAmp placement tool using deep models. Appl Soft Comput 115:108188

17. Gusmão A, Horta N, Lourenço N, Martins R (2021) Late breaking results: attention in Graph2Seq neural networks towards push-button analog IC placement. In: ACM/IEEE design automation conference

18. Andraud M, Stratigopoulos H, Simeu E (2016) One-shot non-intrusive calibration against process variations for analog/RF circuits. IEEE Trans Circ Syst I Regul Pap 63(11):2022–2035

19. Passos F et al (2018) Enhanced systematic design of a voltage controlled oscillator using a two-step optimization methodology. Integr VLSI 63:351–361

20. Mendes L, Vaz J, Passos F, Lourenço N, Martins R (2021) In-depth design space exploration of 26.5-to-29.5-GHz 65-nm CMOS low-noise amplifiers for low-footprint-and-power 5G communications using one-and-two-step design optimization. IEEE Access 9:70353–70368

Chapter 2
State-of-the-Art

2.1 Knowledge-Based Sizing

The first analog IC sizing automation tools were knowledge-based ones, such as IDAC [18] and BLADES [19], that aim to formalize the design process by leveraging design plans obtained via expert knowledge. These analog IC sizing frameworks use circuit equations and predefined design strategies to determine component sizes that meet specific performance targets. Although such methods have demonstrated satisfactory results for automated analog IC sizing, particularly due to their short execution times, the process of developing the underlying design plan is complex and time-consuming. Furthermore, continuous advancements in fabrication technologies and the growing complexity of circuit topologies require these plans to be frequently updated. Nonetheless, since their results typically fall short of simulator-level accuracy, these tools remain primarily suited for early-stage, first-cut design exploration.

2.2 Optimization-Based Sizing

Targeting optimal results, and to address the limitations of knowledge-based sizing, the next generation of automatic analog/radio-frequency (RF) sizing tools apply optimization techniques to the design of these circuits and can be categorized into equation-based or simulation-based evaluation, depending on the method used to evaluate the circuit's performances.

P. A. Oliveira Paiva et al., *Analog Integrated Circuit Design Under PVT Conditions*, SpringerBriefs in Applied Sciences and Technology, https://doi.org/10.1007/978-3-032-19409-1_2

2.2.1 Equation-Based Evaluation

Equation-based optimization approaches rely on analytical design equations to estimate circuit performance. Early tools such as OPASYN [20] and CADICS [21] required manual derivation of these equations, while ISAAC [22] later automated the process by generating simplified design equations automatically. Although these methods enable rapid evaluation and are well-suited for preliminary design stages (analogous to knowledge-based sizing), their reliance on approximations inherent in the equations limits their ability to accurately mapping certain design characteristics, often resulting in reduced overall precision.

2.2.2 Simulation-Based Evaluation

Simulation-based optimization is the predominant method in both academic and industrial domains, given that most designers opt to refrain from the errors induced by the approximations made in equation-based performance evaluation. These methods rely on off-the-shelf circuit simulators to evaluate circuit performance and offer greater generalization capabilities compared to the previously discussed approaches. However, the extensive execution times of SPICE-based circuit evaluations, coupled with the large number of simulations (often in the hundreds or thousands) required to achieve convergence, can result in prohibitively long optimization processes.

In Fig. 2.1, a simulation-based sizing loop is presented. These tools often employ optimization algorithms, particularly evolutionary algorithms, which efficiently navigate the design parameter space by balancing exploration and exploitation to maintain solution diversity. These synthesizers begin with an initial population of dozens or hundreds of individuals (i.e., the parametrized circuit's sizing), each containing input values that will be optimized. To infer their feasibility, each candidate is evaluated through circuit simulations under various testbenches, including transient, AC, and noise analyses, to extract performance metrics. The optimization engine then uses these results to produce the next set of sizing candidates, commonly employing algorithms such as NSGA-III [23] and GDE3 [24]. One example of this approach was presented in [25], where the AIDA environment, an analog IC design automation workflow, implements a design methodology from circuit-level specifications to a physical layout description.

To address some of the research objectives outlined in the previous Chapter, an enhanced version of the simulation-based sizing tool AIDA, AIDA-C [26], was adopted and its use will be explained further in Chap. 4. Various measures from multiple testbenches can be combined and subsequently incorporated into expressions that serve as targets for the constrained multi-objective optimization (MOO) task. Through the combination of a powerful MOO kernel and industry-grade circuit simulators including Siemens EDA's HSPICE, Cadence's SPECTRE or Synopsys' ELDO, the fulfillment of robust design requirements becomes possible. Adding to

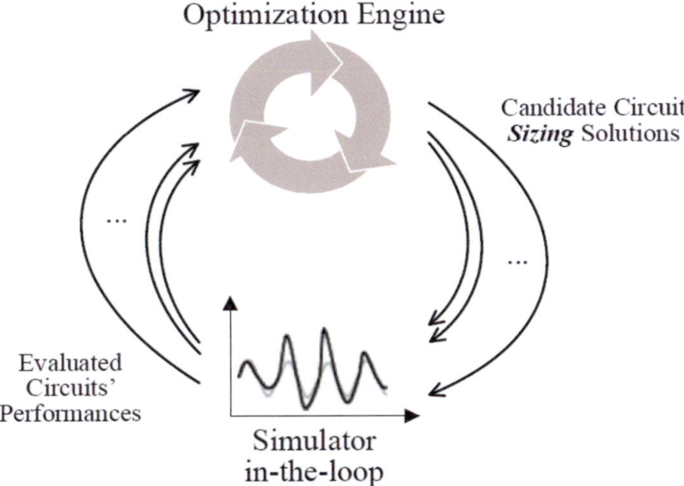

Fig. 2.1 Simulation-based sizing loop. (Reprinted from [27])

these robust requirements, the adoption of a worst-case methodology with PVT corner analysis, where each sizing solution is simulated under various combinations of process, voltage, and temperature deviations, can be incorporated into this loop as well. Despite its importance, this approach remains computationally expensive in modern electronic design automation (EDA) tools, as the simulation costs are considerably higher than those at the nominal (TT) corner, as shown in Fig. 2.2.

Traditional Optimization Flow

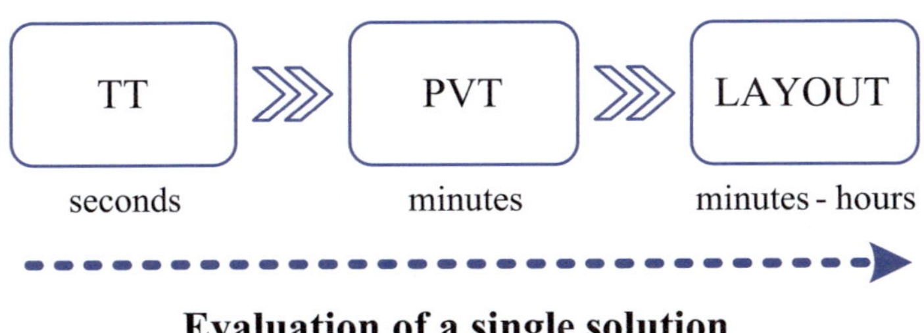

Evaluation of a single solution

Fig. 2.2 Traditional optimization flow of Analog ICs

2.3 Machine Learning in Analog/RF IC Sizing Automation

Since each simulation within an optimization loop can be time-consuming, considerable effort has been devoted to developing techniques that reduce the simulator's workload, a significant amount of which relying on ML algorithms. ML is a branch of artificial intelligence (AI) focused on developing algorithms that can learn from data and has revolutionized many areas of research over the last few decades, from image recognition to natural language processing and time series forecasting. At its core, these techniques seek to approximate complex input–output mappings, uncover hidden structures in data or devise strategies for decision-making without the need of explicit, hard-coded instructions.

Of the different ML techniques that exist, artificial neural networks (ANNs) have been studied extensively to speed-up simulation-based optimizations. ANNs are complex networks comprised of nodes that resemble neurons from the human brain, adjusting the strength of each connection by comparing the desired outputs with obtained ones. The multiple applications of these networks are going to be overviewed in the following sub-sections.

2.3.1 Speeding-Up Simulation-Based Sizing with Artificial Neural Networks

With the growing availability of computational resources, especially through advancements in graphics processing units (or GPUs), the adoption of surrogate models such as ANNs has become a common practice in simulation-based circuit sizing. Thanks to their fast prediction capabilities and ample capacity at handling large amounts of data, these models can be employed to significantly alleviate the simulator's workload in simulation-based sizing loops, aiding it in specific tasks or, in more extreme scenarios, by entirely replacing it, successfully operating as an alternative evaluation engine to off-the-shelf circuit simulation.

ANNs consist of an input and output layer, and one or more hidden layers. However, as demonstrated in [28], an ANN with a single hidden layer provided with a sufficient number of hidden neurons can act as a universal approximator. This enables ANNs to effectively model complex nonlinear input–output relationships, making them particularly well-suited for tasks like analog/RF IC sizing.

Nevertheless, ANNs still have significant drawbacks, like their highly data-dependent nature, requiring large datasets to achieve good generalization to new, unseen inputs, while mitigating the effect of overfitting [29]. Another criticized flaw is ANNs' black-box nature, meaning that the internal logic by which an ANN learns corresponds to an optimization problem involving from thousands to millions of parameters, culminating in a lack of interpretability of the learning process.

While these disadvantages pose significant challenges in the development of ANN-based models, the flexibility and expressive power they offer has continuously driven their integration in end-to-end ML frameworks in the scope of EDA, with applications extending from fault testing [15], modeling [30] and even layout generation [11].

In [1], an ANN-based performance estimation approach was applied to the automatic synthesis of CMOS operational amplifier topologies. To ensure accuracy and efficiency, the ANN models were integrated into a genetic algorithm (GA)-based circuit synthesis framework. Each performance parameter was predicted using an ANN trained on SPICE-generated data, and prediction accuracy was verified by comparing network outputs with simulation results. Despite using shallow architectures (one hidden layer with 8–13 units), the models achieved satisfactory generalization. Compared to the traditional SPICE-based sizing loop, this approach was significantly faster: a 10,000-iteration optimization involving eight performance metrics, requiring roughly 44 h with SPICE, was reduced to just 80 s using the ANN surrogates, achieving a 2000× speed-up. However, since the method depends on pre-collected simulation data, adapting it to circuits with different functionalities would require new SPICE testbenches, limiting its generality.

In [31], a similar approach using ANNs to accelerate simulation-based circuit synthesis was proposed. Unlike the previous work, where networks were trained beforehand, here the sizing loop remains unchanged for a certain number of generations in order to collect accurate simulator data, which is used to train the ANNs online (during the optimization process). By reusing data typically discarded in conventional approaches, it eliminates the need for a pre-existing dataset, allowing application to a broader range of analog circuit topologies without an extensive data acquisition phase. The method was validated through the automatic synthesis of a single-stage amplifier and a folded-cascode operational transconductance amplifier (OTA), where each ANN predicted a single performance metric, mapping 7 and 19-dimensional parameter spaces, depending on the circuit. Depending on the generation chosen for training, speed-ups of up to 64.8% were achieved, though model performance and generalization depended strongly on the quantity and quality of the training data, with earlier training phases leading to less accurate predictions.

Building on the same methodology, [32] employs ANNs to improve the continuity of the Pareto optimal front (POF) in analog IC MOO. Because of randomness and limited population size, the final POF may contain gaps between optimal points. Increasing the number of individuals or generations can mitigate this but greatly raises simulation costs. To address this, the method transitions from multi-objective to single-objective optimization (SOO), while using the coarse POF from the MOO phase as a guide to make the final POF more continuous. To accelerate evaluations, ANNs trained on data from the MOO stage are integrated into the SOO phase. Results from two case studies show that this approach enhances the smoothness and quality of design trade-offs, achieving speed-ups of up to 29.6 × with prediction errors below 1%.

While previous methods employed ANN-based performance regressors only under TT conditions, ensuring industry-grade robustness requires PVT-inclusive optimization, as mentioned previously. In [33], an ML/DL-based sizing methodology specifically designed for PVT-inclusive optimization was introduced for the first time in the literature. Unlike prior frameworks, this approach integrates ANN-based PVT performance regressors within a simulation-based sizing loop to significantly reduce optimization time. Rather than fully replacing the simulator, two control phases are incorporated to mitigate erroneous predictions. The PVT estimator consists of parallel ANNs, each predicting performance values for a specific PVT corner, while the TT corner is always simulated during optimization. This strategy was evaluated on a dual-mode class-C/D voltage-controlled oscillator (VCO) in a TSMC 65 nm CMOS process, achieving a 78.5% reduction in simulation workload and a 2.92× speed-up, saving over 16 days of computation, while maintaining competitive sizing accuracy. However, generating the dataset for the PVT regressors required a full PVT-inclusive optimization of the same circuit. Using a GA employing a population of 256 over 350 generations, this process took approximately 612 h (about 25 days) on an Intel E5-2630-v3@2.40 GHz workstation with eight parallel licenses.

2.3.2 Reinforcement Learning in Analog/RF IC Sizing

RL is a subfield of ML in which an agent (typically an ANN) learns to make decisions by interacting with an environment to achieve a specific goal. The agent performs actions within this environment, receives feedback in the form of rewards or penalties, and adapts its behavior to maximize the cumulative reward over time. This learning process is inspired by behavioral psychology, where actions leading to favorable outcomes are reinforced, while those producing negative outcomes are discouraged.

The fundamental components of RL include the agent, the environment, the states (representing the environment at a specific moment), the actions (choices available to the agent), and the rewards (feedback signals). In essence, when the agent is in a given state ($s \in S$), it selects a valid action ($a \in A$) and receives an immediate reward ($r \in R$) from the environment based on that action. In an analog IC sizing task, the agent selects circuit parameters to be simulated and receives corresponding performance metrics and rewards. The agent's goal is to develop a policy, a strategy defining the optimal action for each state, to maximize long-term rewards. The agent's architecture, which includes the neural network, number of layers, and layer size, as well as the learning algorithm (e.g., policy gradient, Q-learning), must be carefully aligned with the task's complexity. The state should contain just enough information to provide the agent with meaningful context for its decisions. The action space can be defined as either continuous or discrete, depending on the problem formulation. And the reward function must be designed to reflect all relevant aspects of performance, providing clear guidance to the agent during training.

To address the analog and mixed-signal modules challenge of a long human-in-the-middle iteration loop, [34, 35] present AutoCkt, an ML framework that automatically optimizes analog circuits performance using discrete actions within a solution space. It implements a deep reinforcement learning agent with sparse subsampling training and aims to find post-layout circuit parameters for any given target specifications, while gaining knowledge of the entire design space in the same style as a circuit designer.

Beginning with the agent's formulation, the method involves progressing through a sequence of environment interactions, collecting rewards at each step until either the objective is achieved or a fixed step limit is reached. After executing several of these trajectories, the neural network is then updated using a policy-gradient update to increase the expected total reward. The work considers N sizing parameters to be tuned to optimize M design specifications, and every time the environment is reset, the parameters are initialized to the center point. Then, the agent receives a state composed of the observed performance o (obtained through simulation of the circuit), the target specifications o^*, and the current parameters and is allowed to increment, decrement, or keep the same value for each circuit parameter. The agent is given H total simulation steps to reach o^*. If the target is achieved in less than H steps, the trajectory ends.

To train the RL agent, the framework begins by selecting M random target specifications. Each target is created by independently sampling every performance metric within user-defined minimum and maximum bounds, based on the circuit topology and application. Next, L trajectories are generated, each assigned one of these sampled targets. The agent employs a three-layer ANN with 50 neurons per layer and uses proximal policy optimization (PPO) as its learning algorithm. The reward for each trajectory is computed by accumulating the step-wise rewards, as shown in (2.1),

$$r_o = \min\left(\beta \frac{o - o^*}{o + o^*}, 0\right), \quad r = \sum_{o \in O} r_o, \quad R = \begin{cases} r, & if \ \ r < 0 \\ 10 + r, & if \ \ r = 0 \end{cases} \tag{2.1}$$

where β is 1 for metrics being maximized, -1 for metrics being minimized, and $-1 < \beta < 0$ for minimized metrics that are not generally defined as hard constraints. The agent is not rewarded for over satisfying any one given metric, and if the sum of all r_o rewards is zero, the obtained performance achieves the target specification, and the agent receives an additional scalar term of 10.

Once training is complete, the agent is deployed to explore a previously unseen design space. In this phase, layout-related data is considered. Instead of deploying the trained agent in an environment that solely consists of post-layout simulations, the schematic and parasitic distributions are combined by taking steps from both environments. This reduces runtime by avoiding unnecessary, expensive parasitic evaluations while also improving the agent's robustness to layout-induced variations. In practice, after receiving a target specification, the agent first performs schematic simulations until it reaches o^*. The resulting parameters are then evaluated in the parasitic-aware environment. If the specifications are not met under parasitics, the layout performance is fed back to the agent so it can adjust the schematic parameters and attempt again. This loop continues until the target specification is achieved under parasitics or the maximum step count is reached. By combining schematic and layout simulations, the agent progresses through a familiar distribution before confronting the more complex parasitic distribution, thereby improving its understanding and generalization of the new design space.

The approach was tested on 5 different circuit topologies developed in multiple technologies, from 16 to 45 nm, and was able to achieve feasible sizing solutions for all performance requirements, to varying degrees of efficiency, all of them acceptable.

Inspired by multiagent planning theory and hierarchical design work, [36] and [37] propose a multiagent-based RL (MA-RL) framework to tackle the analog circuit design automation issue. The idea is to partition a complex analog circuit into several subblocks based on the circuit topologies and design knowledge. After the partitioning, each subblock is assigned a corresponding RL agent and a design goal. The RL agent tries to optimize towards its own reward while all agents interact with each other to accomplish the overall target. This approach reduces the complexity of the design space and helps ML algorithms to accomplish a more efficiently exhaustive search.

Each agent is composed of an actor network and a critic network. The actor's job is to find the action to optimize the figure of merit (FoM), while the critic informs the actor how good the action is and how it should improve. Critics in different agents interact with each other by shared states and actions of all agents, and then they influence their corresponding actor network by changing their loss function. The system objective is then to maximize the FoM of the entire circuit. In the implementation of the MA-RL framework, both actor and critic are ANNs with one input layer, one output layer and two hidden layers. The input of the actor network is its state, and the output is the action. The input of the critic network is a concatenation of all states and actions, and its output is a Q value.

An action is composed of design parameters, such as device sizes and passives. It is a vector concatenating this information of its tunable devices within its corresponding sub-block, e.g., transistor width (W), length (L), capacitance (C), and Resistance (Res), etc. as can be seen in Eq. (2.2).

$$Action^i\left[W_1^i, L_1^i, W_2^i, L_2^i, \ldots, W_k^i, L_k^i, Res_1^i, Res_2^i, \ldots, C_1^i, C_2^i, \ldots\right] \quad (2.2)$$

The state of each agent is a vector containing the circuit information of the corresponding subblock observed at the moment. In this work, two different state definitions are tested: the first is composed of the simulated operating conditions for a subblock, like g_m, V_{sat}, etc., as shown in (2.3); and the second is composed of the current performance reached in the corresponding subblock as in (2.4), where $Spec_1^i$ is the nth performance Spec of the ith subblock. For example, the state of each subblock is a set of circuit performance, such as gain, power consumption, and so on.

$$State^{i,a} = [gm_1^i, gm_2^i, \ldots, V_{sat}^i, V_{sat}^i, \ldots] \quad (2.3)$$

$$State^{i,b} = [Spec_1^i, Spec_2^i, \ldots, Spec_n^i] \quad (2.4)$$

The reward of each agent R^i include the reward calculated from the performance specification of that subblock R_i and the reward calculated from the performance specification of the entire circuit R_t as shown in (2.5). A more detailed definition of each of these rewards is in [36] and [37].

$$R^i = R_i + R_t \tag{2.5}$$

These works also test three different algorithms of the optimization process, each with different strengths and weaknesses: the multiagent deep deterministic policy gradient (DDPG) (MADDPG), the multiagent twin delayed DDPG (MATD3), and the multiagent PPO (MAPPO). To showcase the capabilities of every MA-RL methodology, three circuits are proposed. The first circuit is a gain boost amplifier designed on commercial 180 nm mixed-signal CMOS process design kit (PDK), composed of 21 design parameters. The second circuit is a delay-locked loop circuit on commercial 130 nm mixed-signal CMOS PDK, composed of 42 design parameters. The third circuit is a successive approximation register designed on commercial 65 nm mixed-signal CMOS PDK, composed of 28 design parameters.

The parameters optimized by the agent include transistor widths and lengths, as well as passive component values. All test cases were simulated using the Synopsys' HSPICE circuit simulator. The MA-RL approaches outperform traditional optimization methods, such as Monte Carlo, GA, and Bayesian optimization (BO), and also surpass a single-agent RL implementation based on DDPG across all three circuits. Among the proposed algorithms, MATD3 and MAPPO deliver the strongest performance: MATD3 achieves the best results on the first and third circuits, while MAPPO performs best on the second.

Another RL approach is RoSE-Opt, introduced in [38] and [39]. RoSE-Opt introduces two key features: first, it embeds essential analog design domain knowledge directly into the learning process, including circuit topology information, PVT variations and post-layout parasitics. Second, it employs a two-level optimization strategy that combines BO and RL to boost sample efficiency. BO serves as a front-end explorer, performing a fast, coarse search to identify a useful suboptimal starting point for the RL agent. This warm start enables the RL agent to reach optimal solutions with fewer interactions, reducing reliance on costly circuit-level simulations.

As mentioned before, the RL backbone is composed of five essential components: the reward, action space, state space, environment and agent. Starting with the reward, at each time step k, the agent implemented receives r_k, defined as:

$$r_k = \text{Mean}\left(\sum_{j=0}^{j=M-1} r_j\right); \text{ if } \exists j \in [0, M-1], \ r_j < 0; \tag{2.6}$$

$$\text{or } r_k = R, \text{ if } \forall j \in [0, M-1], \ r_j = 0$$

where $r_j = \sum_{i=0}^{N-1} w_i \times \min\{(s_i^j - g_i)/(s_i^j + g_i), 0\}$ is the sub reward of the jth corner, calculated by a weighted sum of the normalized difference between ith intermediate circuit specification of the jth corner s_i^j and the ith design goal g_i. All circuit specifications are equally important, i.e., $w_i = 1$. M represents the number of PVT corners, and N indicates the number of circuit specifications. To not overly optimize, r_j has an upper bound of 0. A reward $R = 10$ is given once all specifications meet the design goal under all PVT corners.

For the action space, RoSE-Opt uses a discrete action space (increase, decrease, or keep) to tune device parameters with a predefined fine-grained step size. For the state space, the intermediate circuit specifications across all PVT corners, along with the circuit topology, are included, capturing the most relevant information from the design environment. To better leverage structural information from the circuit itself, a graph representation $G(V, E)$ is used, where each node in V corresponds to a device and each edge in E represents a connection between devices.

For the environment definition, it consists of the netlist of a given analog circuit, a commercial circuit simulator and a data processing module, responsible for dealing with the simulated results and returning the reward to the agent, using (2.6), while also updating the device parameters to rewrite the circuit netlist based on the actions of the agent. And finally, regarding the agent architecture, these works adopt an actor-critic method, integrating three unique features. First, a custom network architecture is employed, using a graph neural network (GNN) and a fully connected neural network (FCNN). Second, dynamic device parameters are used to encode node features, as they ease the process of learning relationships between device parameters and design goals. Finally, a graph attention network (GAT) is chosen as the backbone of the GNN, as the multi head attention mechanism of GAT helps to learn more complex and higher-dimensional interactions between circuit node and its neighbors. The combination of the GAT and the FCNN form the policy network, which are trained using PPO algorithm.

To test the framework for efficacy, robustness and sampling efficiency, the authors adopt four circuits: a single-stage operational amplifier (Op-Amp), a two-stage Op-Amp, a folded-cascode Op-Amp and a three-stage nested Miller compensation Op-Amp with feedforward transconductance stage. The results show that the BO-based jump-start has a definite improvement over utilizing random starting points, as the RL agent without an optimized starting point often needs more circuit-level simulations to achieve the same reward as the one with the optimized starting point. Furthermore, the inclusion of domain knowledge is an asset in testing parasitic-aware device parameter optimization capabilities. Without accounting for parasitic effects during pre-layout design, the resulting device parameters may fail to meet specifications once transferred to layout. The authors address this issue, and experiments with a single-stage Op-Amp indicate that RoSE-Opt typically needs no more than two rounds to produce parameters that also satisfy layout-level specifications.

Finally, without accounting for parasitic effects during pre-layout design, the resulting device parameters may fail to meet specifications once transferred to layout. The method addresses this issue, and experiments with a single-stage Op-Amp indicate that RoSE-Opt typically needs no more than two rounds to produce parameters that also satisfy layout-level specifications. Furthermore, [38] and [39] also compare two different algorithms in PPO and DDPG. When compared against other existing learning and optimization methods, including RL-based sizing implementations, RoSE-Opt shows dominance while incorporating PVT corners.

One last RL implementation should be noted. Kong et al. [40] introduces PVTSizing, a batch-sampling optimization framework for PVT-robust analog circuit synthesis. This implementation is similar to RoSE-Opt, using trust-region Bayesian optimization (TuRBO) to generate high-quality initial datasets and reference points, followed by multi-task RL for PVT optimization. TuRBO is chosen for its scalable global search, achieved through multiple local BO trust regions, which are hyper-cubes that match the search-space dimension and are centered on the best solution found within each region.

The authors define the analog circuit problem as a constraint satisfaction problem:

$$\begin{aligned} minimize \quad & 0 \\ subject\ to \quad & F_i\big(X\,|T_j\big) < C_i, j = 1, 2, \ldots k \end{aligned} \tag{2.7}$$

where:

$$\begin{aligned} X &\in D \\ X &= X_1, X_2, \ldots, X_n \\ D &= D_1, D_2, \ldots, D_n \\ T &= T_1, T_2, \ldots, T_k \\ C &= C_1, C_2, \ldots, C_m \end{aligned} \tag{2.8}$$

X is the n-dimension design solution of the sizing problem corresponding to n device sizing variables, D is the n-dimension design space, C is the m-dimension constraint vector representing m performance metrics, T is the PVT corner, where there are k PVT conditions in total, and F is the mapping from the circuit design solution to performance metric, meaning that $F_i(X\,|T_j)$ is the ith performance metrics when design solution X is simulated under PVT condition T_j.

Regarding the reward and FoM, these are defined to convert multiple objectives into a single target, simplifying the problem:

$$FoM = \begin{cases} r, & r < 0 \\ 0.2, & r \geq 0 \end{cases}$$

$$r = \begin{cases} \sum_{i=1}^{m} \min\left(\frac{F_i - C_i}{\max(F_i + C_i, C_i)}, 0\right), & if\ we\ want\ F_i \geq C_i \\ \sum_{i=1}^{m} \min\left(\frac{C_i - F_i}{\max(F_i + C_i, C_i)}, 0\right), & if\ we\ want\ F_i \leq C_i \end{cases} \tag{2.9}$$

Initially, the framework runs TuRBO for initial sampling, generating design solutions that meet the constraints under nominal conditions. Then, these design solutions are sampled under full PVT conditions to generate the initial dataset. Then, each iteration of the RL-loop can be divided into six steps. First, the PVT corners are pruned to create a reduced corner set by applying the K-means algorithm to the last sample of each corner and selecting the worst corner from each of the B clusters. Second, this reduced set is fed to the actor network to produce B candidate design solutions. Third, the critic network filters these solutions by predicting which ones are most likely to accelerate optimization. If none qualify, the process skips to the final step. Fourth, the selected solutions are evaluated under the reduced corner conditions and the buffers are updated. Fifth, if the latest buffered performances meet all constraints, a full PVT test is performed. If all targets are satisfied, optimization stops; otherwise, the buffers are updated again. Finally, the RL agent is trained using only the data corresponding to the reduced corner set.

The multi-task RL agent, based on the actor-critic method, handles the optimization conflict among corners. The actor includes a 5-layer neural network. It receives as input a 7-dimensional vector, where the first five dimensions represent process variation by one-hot encoding and the last two dimensions are for normalized voltage and temperature, and outputs a n-dimensional normalized design solution. The critic also includes a 5-layer neural network. It receives as input a n-dimensional design solution and outputs a k-dimensional vector representing the predicted FoM of input under each PVT condition. During the update, each task is defined as searching for a design solution that meets constraints under a certain PVT condition.

Because each iteration produces B candidate solutions but only one is sampled, another pruning strategy is required. Instead of relying on absolute FoM predictions, the critic estimates the relative quality between candidates, achieving about 66% accuracy, which is sufficient for pruning. As the framework tracks the last sample of each corner, sampling a solution X may improve some corners and worsen others. The "improvement number" of X is defined as the difference between the "up number" and the "down number". The definition of the "up number" of a certain X is the number of corners being better after sampling X, while "down number" is the number of corners being worse. In each iteration, the critic predicts the "up number" for every candidate and randomly selects one whose value exceeds a sampling threshold β, which is updated each iteration. Faster convergence is achieved by preferring solutions with higher improvement numbers, as shown in (2.10), where P_{inc} and P_a represent the current quality of the actor and critic. Sweeping β from 0 to B identifies the threshold that yields the best improvement rate.

$$G(\beta, P_{inc}, P_a) = \textit{Improvement speed} = \frac{E(\textit{Improvement number})}{E(\textit{Sample interval})} \tag{2.10}$$

$$\beta = \textit{argmax}(G(\beta, P_{inc}, P_a)) \tag{2.11}$$

To demonstrate the framework's capabilities, four circuits were chosen, being a folded-cascode OTA in TSMC 180 nm technology, with 20 design variables and 30 PVT corners, a strongArm latch in TSMC 28 nm technology, with 14 design variables and 30 PVT corners, a bandgap voltage reference in TSMC 180 nm technology, with 8 design variables and 15 PVT corners, and a floating inverter pre-amplifier in TSMC 180 nm technology, with 6 design variables and 30 PVT corners. Among all four testcases, PVTSizing achieves the highest sample efficiency and lowest runtime, when compared to other works. Furthermore, an ablation study was conducted that showed that the integration of TuRBO and the zoom target metrics mechanism contributed to the sample efficiency and success rate. The usage of the critic-assisted pruning also proved to improve the success rate but to reduce the sample efficiency in the last circuit.

2.3.3 Transfer Learning in Analog/RF IC Design

Transfer Learning is a ML paradigm in which knowledge acquired from training a model on a source task within one domain is leveraged to improve learning on a different but related target task. Instead of building a new model from scratch, it reuses the representations learned from the source task to accelerate convergence and enhance generalization on the target task. This is particularly advantageous when labeled data for the target problem is limited or expensive to generate, as in PVT-inclusive analog IC optimizations, where obtaining simulator-accurate data is computationally demanding due to high simulation costs and long optimization runtimes. In practice, a model is first trained on a large source dataset, and its learned parameters, particularly the weights of an ANN, are transferred to initialize a target model. The lower layers, which capture domain-invariant features, are typically frozen to retain their parameters, while the upper layers, which encode task-specific information, are fine-tuned or replaced to meet the requirements of the new task [41]. This transfer of knowledge significantly reduces training time and computational effort while improving robustness and performance in low-data settings.

Some works have explored transfer learning–based approaches for automating analog/RF IC design. In [42], an analog/mixed-signal circuit optimization framework is introduced that integrates RL with GNNs for both schematic and post-layout optimization. GNNs serve as RL function approximators to enhance convergence and generalization, while transfer learning enables adaptation across different technology nodes and circuit topologies. In [43], a global mapping model fusion (GMMF) technique is proposed to efficiently construct post-layout performance models by reusing schematic-level neural network models. Instead of training new models from scratch, GMMF introduces a mapping layer that links schematic and post-layout metrics through a small set of coefficients optimized via differential evolution, significantly reducing the number of required post-layout simulations while preserving high accuracy. Likewise, the work in [43] presents a fine-tuning-based model fusion strategy for post-layout performance modeling. In this method, a pre-trained schematic-level ANN is fine-tuned using a limited number of post-layout samples, leveraging its learned parameters to accelerate training. Experiments across multiple analog circuits showed more than a $9\times$ reduction in required samples compared to conventional approaches.

2.4 Conclusions

In this Chapter, various methods used in circuit optimization and sizing were introduced and compared based on their evaluation engines, with particular emphasis on ML-based approaches. The most widely accepted engines in the industry remain simulation-based implementations, offering strong generality and providing an easy-to-use and accurate modeling framework. However, their main drawback are the potentially long runtime, especially as SPICE-like simulation times grow and with the added burden of optimizing sizing while taking into account PVT corners for robust analog IC design. Recent work attempts to mitigate this issue by incorporating ML technology into the optimization process, either by replacing the circuit simulator or by complementing it.

Focus is given to two training methodologies: RL and transfer learning. RL-based implementations have already been studied in the past, but the incorporation of PVT corners into the optimization problem is a recent one. Transfer Learning solutions have mostly been targeted to sizing processes under nominal operating conditions and post-layout performance prediction, although it seems to be an adequate solution for the PVT-aware optimization, given the lack of available data in PVT corners.

References

1. Wolfe G, Vemuri R (2003) Extraction and use of neural network models in automated synthesis of operational amplifiers. IEEE Trans Comput Aided Des Integr Circ Syst 22(2):198–212
2. Alpaydin G, Balkir S, Dundar G (2003) An evolutionary approach to automatic synthesis of high-performance analog integrated circuits. IEEE Trans Evol Comput 7(3):240–252
3. Liu H, Singhee A, Rutenbar R, Carley L (2002) Remembrance of circuits past: macromodeling by data mining in large analog design spaces. In: Proceedings 2002 design automation conference, pp 437–442
4. Mina R, Jabbour C, Sakr G (2022) A review of machine learning techniques in analog integrated circuit design automation. Electron MDPI 11(3):435
5. Martins R, Lourenço N (2023) Analog integrated circuit routing techniques: an extensive review. IEEE Access 11:35965–35983
6. Wang C, Yang F, Zhu K (2024) AI-enabled layout automation for analog and RF IC: current status and future directions. In: Proceedings of the IEEE international symposium on radio-frequency integration technology, pp 1–3
7. Maji S, Budak A, Poddar S, Pan D (2024) Toward end-to-end analog design automation with ml and data-driven approaches (invited paper). In: Proceedings of the 29th Asia and South Pacific design automation conference, pp 657–664
8. Martins R (2025) A survey of machine and deep learning techniques in analog integrated circuit layout synthesis. Microelectronics MDPI 1(1, 2)
9. Azevedo F, Lourenço N, Martins R (2025) Comprehensive application of denoising diffusion probabilistic models towards the automation of analog integrated circuit sizing. Expert Syst Appl 290:128414
10. Eid P, Azevedo F, Lourenço N, Martins R (2025) Using denoising diffusion probabilistic models to solve the inverse sizing problem of analog integrated circuits. AEU Int J Electron Commun 195:155767

11. Zhu K et al (2019) Genius route: a new analog routing paradigm using generative neural network guidance. In: Proceedings of the IEEE/ACM international conference on computer-aided design, pp 1–8
12. Gusmão A, Horta N, Lourenço N, Martins R (2022) Scalable and order invariant analog integrated circuit placement with attention-based graph-to-sequence deep models. Expert Syst Appl 207:117954
13. Gusmão A, Póvoa R, Horta N, Lourenço N, Martins R (2022) DeepPlacer: a custom integrated OpAmp placement tool using deep models. Appl Soft Comput 115:108188
14. Gusmão A, Horta N, Lourenço N, Martins R (2021) Late breaking results: attention in Graph2Seq neural networks towards push-button analog IC placement. In: ACM/IEEE design automation conference, pp 1360–1361
15. Andraud M, Stratigopoulos H, Simeu E (2016) One-shot non-intrusive calibration against process variations for analog/RF circuits. IEEE Trans Circ Syst I Regul Pap 63(11):2022–2035
16. Passos F et al (2018) Enhanced systematic design of a voltage controlled oscillator using a two-step optimization methodology. Integr VLSI 63:351–361
17. Mendes L, Vaz J, Passos F, Lourenço N, Martins R (2021) In-depth design space exploration of 26.5-to-29.5-GHz 65-nm CMOS low-noise amplifiers for low-footprint-and-power 5G communications using one-and-two-step design optimization. IEEE Access 9:70353–70368
18. Degrauwe MGR et al (1987) IDAC: an interactive design tool for analog CMOS circuits. IEEE J Solid-State Circ 22(6):1106–1116
19. El-Turky F, Perry EE (1989) BLADES: an artificial intelligence approach to analog circuit design. IEEE Trans Comput Aided Des Integr Circ Syst 8(6):680–692
20. Koh HY, Sequin CH, Gray PR (1990) OPASYN: a compiler for CMOS operational amplifiers. IEEE Trans Comput Aided Des Integr Circuits Syst 9(2):113–125
21. Jusuf G, Gray PR, Sangiovanni-Vincentelli AL (1990) CADICS-cyclic analog-to-digital converter synthesis. In: IEEE international conference on computer-aided design. Digest of technical papers, pp 286–289
22. Gielen G, Walscharts H, Sansen W (1989) ISAAC: a symbolic simulator for analog integrated circuits. IEEE J Solid-State Circ 24(6):1587–1597
23. Deb K, Jain H (2014) An evolutionary Many-Objective optimization algorithm using reference-point-based non-dominated sorting approach, part i: solving problems with box constraints. IEEE Trans Evol Comput 18(4):577–601
24. Kukkonen S, Lampinen J (2025) GDE3: the third evolution step of generalized differential evolution. In: 2005 IEEE congress on evolutionary computation, vol 1, pp 443–450
25. Martins R, Lourenço N, Rodrigues S, Guilherme J, Horta N (2012) AIDA: automated analog IC design flow from circuit level to layout. In: 2012 International conference on synthesis modeling, analysis and simulation methods and applications to circuit design (SMACD). Seville, Spain, pp 29–32
26. Martins R et al (2020) Design of a 4.2-to-5.1 GHz ultralow-power complementary class-B/C hybrid-mode VCO in 65-nm CMOS fully supported by EDA tools. IEEE Trans Circ Syst I: Reg Pap 67(11):3965–3977
27. Domingues JL et al (2023) Speeding-up radio-frequency integrated circuit sizing with neural networks. In: Springer briefs in applied sciences and technology, 1st edn. Springer
28. Hornik K, Stinchcombe M, White H (1989) Multilayer feedforward networks are universal approximators. Neural Netw 2(5):359–366
29. Hastie T, Tibshirani R, Friedman J (2009) The elements of statistical learning: data mining, inference, and prediction, 2nd edn
30. Suissa A et al (2010) Empirical method based on neural networks for analog power modeling. IEEE Trans Comput Aided Des Integr Circ Syst 29(5):839–844
31. Islamoglu G, Çaklcl TO, Afacan E, Dundar G (2019) Artificial neural network assisted analog IC sizing tool. In: 16th international conference on synthesis, modeling, analysis and simulation methods and applications to circuit design (SMACD), Lausanne, Switzerland, pp 9–12
32. Çakıcı TO, İslamoğlu G, Güzelhan ŞN, Afacan E, Dündar G (2020) Improving POF quality in multi objective optimization of analog ICs via deep learning. In: 2020 European conference on circuit theory and design (ECCTD), Sofia, Bulgaria, pp 1–4

33. Vaz P, Gusmão A, Horta N, Lourenço N, Martins R (2022) Speeding-up complex RF IC sizing optimizations with a process, voltage and temperature corner performance estimator based on ANNs. In: 2022 IEEE international symposium on circuits and systems (ISCAS). IEEE, pp 1570–1574

34. Settaluri K, Haj-Ali A, Huang Q, Hakhamaneshi K, Nikolic B (2020) AutoCkt: deep reinforcement learning of analog circuit designs. In: 2020 design, automation and test in Europe conference and exhibition (DATE), pp 490–495

35. Settaluri K et al (2022) Automated design of analog circuits using reinforcement learning. IEEE Trans Comput Aided Des Integr Circ Syst 41(9):2794–2807

36. Zhang J, Bao J, Huang Z, Zeng X, Lu Y (2023) Automated design of complex analog circuits with multiagent based reinforcement learning. In: 2023 60th ACM/IEEE design automation conference (DAC), pp 1–6

37. Bao J et al (2024) Multiagent based reinforcement learning (MA-RL): an automated designer for complex analog circuits. IEEE Trans Comput Aided Des Integr Circ Syst 43(12):4398–4411

38. Gao J, Cao W, Zhang X (2023) RoSE: robust analog circuit parameter optimization with sampling-efficient reinforcement learning. In: 2023 60th ACM/IEEE design automation conference (DAC), pp 1–6

39. Cao W et al (2025) RoSE-opt: robust and efficient analog circuit parameter optimization with knowledge-infused reinforcement learning. IEEE Trans Comput Aided Des Integr Circ Syst 44:627–640

40. Kong Z et al (2024) PVTSizing: a TuRBO-RL-based batch-sampling optimization framework for PVT-robust analog circuit synthesis. In: 2024 61th ACM/IEEE design automation conference (DAC), pp 1–6

41. Géron A (2022) Hands-on machine learning with Scikit-learn, Keras, and TensorFlow: concepts, tools, and techniques to build intelligent systems, 3rd edn

42. Li Z, Carusone AC (2024) An open-source AMS circuit optimization framework based on reinforcement learning—from specifications to layouts. IEEE Access 12

43. Wang Z et al (2022) Building post-layout performance model of analog/RF circuits by fine-tuning technique. In: Proceedings of the international symposium on quality electronic design (ISQED). IEEE, pp 1–6

Chapter 3
PVT Corner Conditions Within Reinforcement Learning-Based Analog IC Sizing

3.1 Contributions

Real-world applications demand IC blocks' robustness, and to ensure it, automatic sizing of analog ICs has moved towards exhaustive PVT-inclusive optimizations [20, 21]. In such scenarios, the functional behaviour of each candidate sizing solution is found by simulating the circuit under different operating modes as well as fabrication dispersions and voltage/temperature variations, escalating the time required for optimization. However, as overviewed in Chap. 2 of this book, while RL became, in the last years, an important mechanism on IC sizing optimization, the inclusion of PVT corners in these is still scarcely reported in automatic analog IC sizing literature.

Focusing on RL, recent years have seen various approaches for handling nominal conditions being proposed. These can vary from, for example, single agent [16] to multi-agent [17], usually with the aid of deep neural networks. [18, 19] are the only works in the literature that, very recently, have report PVT conditions. Each presents innovative techniques to handle the complex integration of these conditions, as the problem complexity scales rapidly, and the number of simulations and optimization time increase exponentially. In [18], Bayesian optimization (BO) and RL are combined to achieve sampling efficiency and robust sizing. The proposed framework employs a BO vanguard and a RL backbone. The BO vanguard will first coarsely seek an optimized starting search point that will be given to the RL agent. By doing this, the agent can be trained with fewer interactions with time ineffective circuit-level simulations, improving sampling efficiency. To incorporate corners, that work adopted a brute force approach, as the RL agent simulates every corner at each time step, while employing a proximal policy optimization (PPO) [22] agent. In [19], trust region BO (TuRBO) [23] is used for initial sampling and multi-task RL agents are used for PVT-phase optimization. The proposed framework starts with TuRBO generating design solutions meeting constraints under nominal conditions. Then, these design solutions are sampled under full PVT conditions to generate the initial dataset, which will be updated to replay buffers to save simulation data.

© The Author(s), under exclusive license to Springer Nature Switzerland AG 2026
P. A. Oliveira Paiva et al., *Analog Integrated Circuit Design Under PVT Conditions*,
SpringerBriefs in Applied Sciences and Technology,
https://doi.org/10.1007/978-3-032-19409-1_3

Even though both works leverage BO and RL, each focused on a different approach. In [18], the RL agent starts from a fixed point, obtained from BO, and is learning a path from this point to a solution, without employing any strategy in terms of corner inclusion. In [19], the RL agent is actively trying to generate new possible solutions from previous knowledge, employing a reduced corner strategy to predict if the candidate solution is worth simulating by predicting the potential improvement. Since no other works have discussed different strategies for PVT conditions inclusion in RL, it is mandatory to understand how much information the agent needs to be able to transfer the knowledge from nominal conditions to PVT conditions, while minimizing the number of simulations done, as this is the most time-consuming and critical step. Therefore, this work presents a comparative study of five different approaches for integrating PVT conditions within state-of-the-art RL-driven sizing methodologies. Each approach begins with the same conditions to ensure a fair comparison, and are evaluated within the same circuit topology, comparing agent steps, number of simulations, execution time and final sizing functional behavior. Results reveal the trade-offs of each approach, and the best performing one for this problem is highlighted and discussed to facilitate more research activities within this field.

3.2 Environment Overview

This subsection addresses important aspects related to the environment where the agent is trained. The environment consists of three parts: a circuit topology, a circuit simulator, and a data processing unit that bridges the simulator to the agent. As in [18], the data processing unit is responsible for receiving the agents actions, updating the circuit parameters, simulating the circuit, reading the results of the simulation, and providing the feedback to the agent in the form of the new state of the circuit and the reward attributed to the action that was provided. The environment is set up so that the agent starts with a fixed set of design parameters and its goal is to find a path from this set of design parameters to a solution that achieves the target specifications. It was implemented following the guidelines of Farama Foundation's Gymnasium [24].

3.2.1 Action Space

In this Chapter, a discrete action space is adopted. For each circuit parameter, the agent is allowed three possible actions: increase, decrease, or keep the current parameter value. This corresponds to a multi-discrete action space, where the agent outputs a vector of actions, one per parameter, at each step. Even though discrete actions limit fine-grained control over each parameter, they can simplify greatly the learning process and provide improved training stability. To reduce the impact of the discrete action space, a fixed and appropriate step size is chosen, as too coarse steps can miss optimal design solutions and too fine can slow convergence.

To encourage sample efficiency and minimize the number of circuit simulations, a maximum number of steps per episode is imposed (128 steps). If the agent is not able to find a valid solution within this limit, the environment will reset to the starting point, allowing the agent to explore alternative strategies in subsequent episodes.

At each step, the environment is expecting a vector from the agent composed of one action for each parameter. These actions will be applied to the previous parameter values that are stored inside the environment. The environment is responsible for guaranteeing that all parameters stay within the predefined bounds, only applying the actions that respect them. When a parameter is already at the, assuming, upper bound, and the agent tries to increase it, the environment will keep the value at the upper bound and the resulting state will reflect that no change occurred to that parameter.

3.2.2 Observation Space

The observation space, also known as state, is one of the most important sources of feedback that the agent receives from the environment. As previously stated, the information present in it should be carefully chosen, as too little information can lead to an uninformed or ineffective agent and too much information can overwhelm the agent and slow the learning process due to the increased dimensionality and noise. Picking an appropriate and informative set of variables to properly represent the environment can be tricky but it is an essential task in RL development.

Considering the implementations overviewed in Chap. 2 of this book, the chosen observation space consists of the normalized circuit specifications at the nominal and PVT corners, followed by the current design parameter values, represented by their discrete indices, this is:

$$State = [Specs_{TT}, Specs_{PVT1}, \ldots, Specs_{PVTn}, Parameter\,values] \qquad (3.1)$$

Each specification is normalized to the range $[-1, 1]$. When a circuit simulation arrives as invalid, all the corresponding corner specifications are set to -1, signaling the agent to what occurred. When some of the corners are intentionally not simulated, all their specifications are set to -1.5, which lies outside of the normalization range, allowing the agent to differentiate from an invalid simulation. By adopting this representation, the agent can learn correlations between different PVT corners, identifying in which is hardest/easiest to fulfil the current set of circuit specifications. At the same time, the agent can check the validity of its actions, learning each parameter's bounds and avoiding pointless actions.

3.2.3 Reward Functions

The reward function is a key element of the environment and one of the most important feedback signals that the agent receives. While the observation describes the state of the environment, the reward function quantifies how likely is the given state to reach the design goal, guiding the agent's learning process.

From the works analyzed in Chap. 2 of this book, two of the most common reward design strategies are the formulation of a figure of merit (FoM), which captures performance trade-offs based on selected specs, or a composite reward that explicitly includes all target specifications, often shaped to reflect specification satisfaction or violation.

In the following subsections, the five different reward formulations tested in this Chapter are described, namely: a binary reward; a steps reward; the reward function from AutoCkt [25]; the reward function from PVTSizing [19]; and, a custom FoM-based reward. Each one is analyzed based on the expected results and potential limitations.

3.2.3.1 Binary Reward

As indicated by the name, this reward function, presented in (3.2), consists of a binary signal set to 0 when all specs are met and -1 otherwise. This reward function provides very limited information to the agent, as it does not have a clear way to distinguish from an almost valid and a completely invalid parameter set. While it is not expected that this reward function achieve a satisfactory performance, it is mainly tested to prove the importance of the information that the reward function contains.

$$R = \begin{cases} 0, & \text{if specifications are satisfied} \\ -1, & \text{otherwise} \end{cases} \tag{3.2}$$

3.2.3.2 Steps Reward

This reward function, shown in (3.3), represents a slight improvement over the binary reward. It consists of using the binary signal for each individual specification and then computing the average across all m specifications.

$$R = \frac{1}{m} \sum_{i=1}^{m} r_i, \text{ with } r_i = \begin{cases} 0, & \text{if specification } i \text{ is satisfied} \\ -1 & \text{otherwise} \end{cases} \tag{3.3}$$

This reward function provides more information to the agent, as it is now possible to distinguish how many specifications are still failing the target, but it does not provide any particular information on how close a missing performance is from achieving the target. It is expected that this reward function will help in the likelihood of the agent converging prematurely to a suboptimal solution by offering denser feedback, but the training process is still expected to be difficult due to the sparse and discontinuous nature of the reward.

3.2.3.3 AutoCkt Reward

This reward function, taken from [16] and shown in (3.4), represents a more refined approach compared to the previous reward strategies. It is designed to guide the agent more effectively by quantifying how far the current performance is from meeting the target specifications.

$$
R = \begin{cases} r, & \text{if } r < 0 \\ 10, & \text{if } r \geq 0 \end{cases}, r = \begin{cases} \sum_{i=1}^{m} \min\left(\frac{F_i - T_i}{F_i + T_i}, 0\right), & \text{if we want } F_i \geq T_i \\ \sum_{i=1}^{m} \min\left(\frac{T_i - F_i}{F_i + T_i}, 0\right), & \text{if we want } F_i \leq T_i \end{cases} \tag{3.4}
$$

where F_i is the current value of the ith specification, T_i is the target value, and m is the total number of specifications. This reward penalizes the agent proportionally when the specification target is not met, while giving a large fixed positive reward once all specifications are satisfied. It is expected for this reward function to significantly improve the learning efficiency, compared to the previous reward functions, as it manages to provide more quantitative feedback to the agent.

3.2.3.4 PVTSizing Reward

This reward function, taken from [19] and shown in (3.5), is very similar to the AutoCkt reward function with two key differences: the introduction of a maximum operation in the denominator to ensure positivity and numerical stability, and a smaller constant reward when all specification are met.

$$
R = \begin{cases} r, & \text{if } r < 0 \\ 0.2, & \text{if } r \geq 0 \end{cases}, r = \begin{cases} \sum_{i=1}^{m} \min\left(\frac{F_i - T_i}{\max(F_i + T_i, T_i)}, 0\right), & \text{if we want } F_i \geq T_i \\ \sum_{i=1}^{m} \min\left(\frac{T_i - F_i}{\max(F_i + T_i, T_i)}, 0\right), & \text{if we want } F_i \leq T_i \end{cases}
$$

$$\tag{3.5}$$

where F_i is the current value of the ith specification, T_i is the target value, and m is the total number of specifications. The use of the maximum operation in the denominator ensures that the reward remains well-defined and bounded, even for small or zero-valued specifications. The constant reward of 0.2 gives the agent a smaller stimulus, possibly encouraging the agent to continue exploration instead of converging prematurely to a suboptimal solution.

3.2.3.5   FoM-Based Reward

FoM-based reward functions offer a flexible and frequently used strategy in analog IC sizing optimization. They are typically designed to emphasize key performance trade-offs or to prioritize specific design goals such as power efficiency, speed, gain, or area. As a result, this reward is expected to be able to lead the agent to find better performing values for the specifications that are accounted in the formulation used.

However, one disadvantage is that these reward functions are inherently circuit-dependent, in contrast to the previously proposed functions. Their effectiveness relies heavily on expert knowledge, and poorly chosen FoMs can bias the agent toward suboptimal regions of the design space by overemphasizing some specs while neglecting others. Still, the main goal of this approach is to evaluate whether a domain-informed scalar reward can effectively guide the agent toward high-quality solutions, and how its performance compares to more general, specification-based, reward formulations.

3.3   PVT-Inclusive Approaches

The following subsections detail each approach for PVT-inclusion and present the expected advantages and disadvantages of each.

3.3.1   Brute Force Approach

As suggested by the name, this approach consists of simulating every corner from the start, as shown in Fig. 3.1. Therefore, it is expected to have poor simulation efficiency. Considering an agent that takes, for example, 30 steps to reach the first candidate solution, and each steps costs 5 simulations, it will cost a total of 150 simulations where, most-likely, more than one corner is far from achieving the targets but is still being simulated. The main interest in this brute force is to use it as a benchmark against the other PVT-inclusive approeaches.

3.3.2   Progressive Overload Approach

This approach consists of progressively overloading the environment by gradually introducing more corners. A representation is shown in Fig. 3.2.

First, all corners are ranked by difficulty using:

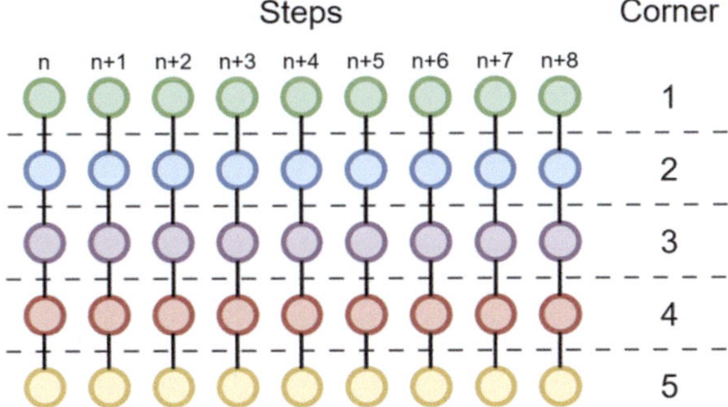

Fig. 3.1 Example of the simulations made using a brute force approach

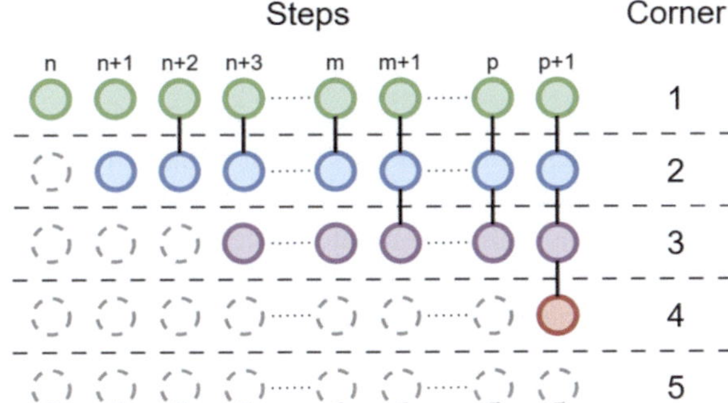

Fig. 3.2 Example of the simulations made using a progressive overload approach

$$Difficulty(C_i) = Npassed + \sum_{j=1}^{m} Fnorm_j \tag{3.6}$$

where C_i is the ith corner, $Npassed$ is the number of targets that are achieved by this corner, and, $Fnorm_j$ is the normalized value of the jth specification from the corner, considering m total specs. Then, the agent starts training only with the hardest corner, i.e., the one furthest from feasibility. In this first stage, every step the agent takes will only cost one simulation, as there is only one corner being considered. When the agent achieves proficiency in the hardest corner, the second hardest corner is added. This means that now every step costs two simulations. As the agent manages to reach proficiency at each level, the environment adds another corner, until there are no corners left. This means that the agent only finds actual solutions when all corners are being simulated at each step. The goal of this approach is for the agent to spend more time gathering information on the hardest corners and progressively use this knowledge as new corner conditions arrive. Just like the Brute Force approach, this approach can still suffer from pointless simulations being run, but at least a considerable speed up is expected.

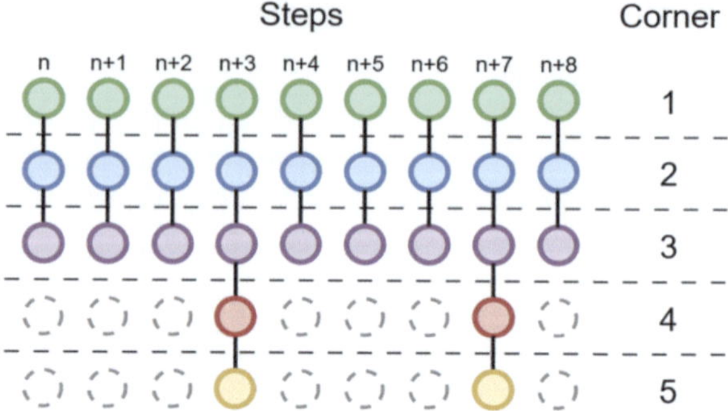

Fig. 3.3 Example of the simulations made using reduced corner set approach

3.3.3 Reduced Corner Set Approach

Inspired by [19] and represented in Fig. 3.3, the goal of this approach is to reduce the full corner set to a smaller number that is representative of it. To achieve it, the K-means algorithm is used to create B corner clusters and only the hardest corner of each cluster will be simulated, effectively cutting simulations. Once a solution is found on the reduced corner set, a simulation of the full corner set is performed to ensure that the given solution is valid.

3.3.4 Scheduler Approach

This approach is an adaptation of [26] and is represented in Fig. 3.4. Instead of simulating a fixed set of corners at each step, the corners are added to the simulation if the candidate solution can achieve the targets in all previously simulated corners. First, the corners are ranked by difficulty using (3.6), so the agent starts looking for solutions in the hardest corner. If the current step does not meet the target in the hardest corner, there is no need to simulate any more corners, and the agent is allowed to take the next step. If the current step meets the target, the second hardest corner is also simulated. If this also meets the target, the third corner will be added, and this behavior is repeated until either one corner does not meet the targets, or all corners meet the targets. When one corner fails the targets, the agent takes a step back, and the environment returns to simulating just the hardest corner. It is considered that a solution was found when all corners are successfully chained and meet the target specifications.

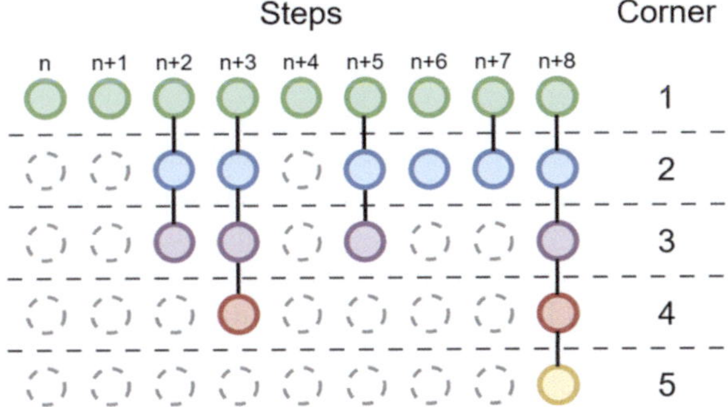

Fig. 3.4 Example of the simulations made using scheduler approach

3.3.5 *Worst-Case Approach*

This approach, represented in Fig. 3.5, is a particular case of the reduced corner set, where the environment simulates the nominal conditions and the worst performing corner of each variation type, being one corner from process, one from temperature and one from voltage, making a total of four simulations at each step. Once a solution that fulfils the specifications in these four corners is found, a full corner simulation is made to check if the solution is valid in the remaining ones.

3.4 Agent Overview

This section delves into the agent, which will encompasses a purely RL approach. First, the chosen agent algorithm is presented, and then, agent implementation details are provided.

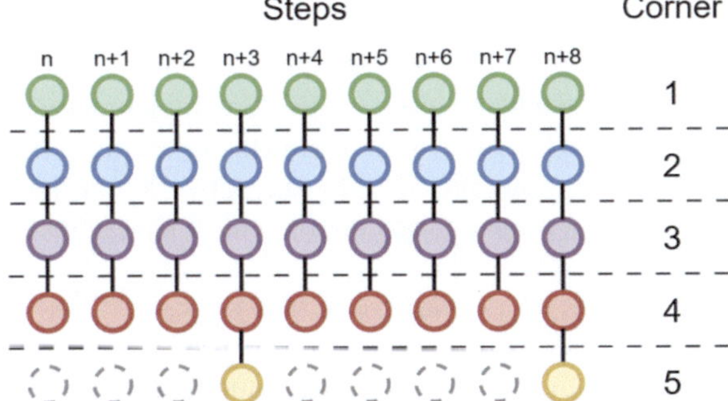

Fig. 3.5 Example of the simulations made using worst-case approach

3.4.1 Agent Algorithm

The algorithm chosen for the feedback loop of this work's agent is PPO. In [18] a study was conducted between PPO and deep deterministic policy gradient (DDPG), which concluded that PPO with a discrete action space demonstrates superior training sample efficiency and more consistent results compared to its continuous counterpart, and also, DDPG with a continuous action space.

PPO is a policy-based RL algorithm that uses gradient ascent to maximize the expected advantage $\widehat{A}_t$ of a certain action a_t in a given state s_t. This is balanced by the *log* probability of taking that action, given by:

$$L(\theta) = E_t\left[\log\pi_\theta(a_t|s_t)\widehat{A}_t\right] \tag{3.7}$$

However, this can lead to unstable learning, as the policy may change too drastically between updates. To address this, PPO introduces a clipped surrogate objective, which limits how large an update to the network should be. First, a ratio is calculated between the probability of taking an action under the new policy and the old policy used to generate the data:

$$r_t(\theta) = \frac{\pi_\theta(a_t|s_t)}{\pi_{\theta_{old}}(a_t|s_t)} \tag{3.8}$$

Instead of only relying on the log-probability, PPO uses this ratio to adjust the objective, while making sure the updates do not jump too far from the original policy. The clipped objective function will then be given by:

$$L^{clip}(\theta) = \widehat{E}_t\left[\min\left(r_t(\theta)\widehat{A}_t, clip(r_t(\theta), 1-\epsilon, 1+\epsilon)\widehat{A}_t\right)\right] \tag{3.9}$$

This approach ensured that no updates are overly aggressive: when $\widehat{A}_t > 0$, the update is clipped to prevent overshooting; when $\widehat{A}_t < 0$, the update is clipped to prevent the policy from degrading too quickly. The advantage can be estimated by using general advantage estimation or simpler methods like Monte-Carlo estimation. The final loss function, used to train the PPO agent, also includes a value loss function between the value provided by the network and the real return expected from that state, as well as an entropy coefficient, S, that helps ensure exploration:

$$Loss = L^{clip}(\theta) - k_1 * L^{Value} + k_2 * S(s_t) \tag{3.10}$$

where k_1 is the coefficient of the value function, which balances how important are the value function predictions, and k_2 is the coefficient of the entropy, which balances the exploration and exploitation trade-off.

3.4.2 *Implementation Details*

The overall workflow of the agent is the following: first, an agent is trained considering only the nominal conditions. The starting point of this agent was chosen to be the middle of the discrete space of every parameter, and all solutions found during training are stored. This agent will be used to compare all the reward functions previously established, and, the best performing agent will be selected to compare the different PVT approaches.

Once the best performing nominal agent is selected, all of its collected solutions will be ranked in full corner conditions. This ranking is done using (3.6) to score the difficulty of each corner and the total score is the sum of all corner difficulties. The solution with the best score will be set as the starting point for the PVT-aware RL agent, which will compare every PVT approach under study. After loading the network of the nominal condition agent, the PVT condition agent starts training, following the approach currently selected.

For both stages of training, the ending condition will be called reaching proficiency, which means that the agent managed to successfully size the circuit in ten consecutive episodes. To compare the reward functions the total number of steps, execution time, and best performing solution will be used. Similarly, to compare the PVT inclusive approaches the total number of steps, total number of simulations, execution time, and best performing solutions for the nominal corner are used. To ensure robustness against variations in random seed, all training sessions are run ten times and the results presented are averages of these runs, unless specified.

Both agents use the PPO algorithm, and their neural network is represented in Fig. 3.6. Instead of employing a shared network for the Actor and the Critic, two different networks are employed. The input layer (a) is the same for both networks. It has a size corresponding to the state of the circuit, comprised of the normalized specifications for each corner and the current circuit parameters. Then, three hidden layers are used for the Actor network (b)–(d), and three hidden layers are used for the Critic network (f)–(h), all of them with 64 neurons each. Finally, the output layer of the Actor (e) are the actions to be applied to the parameters of the circuit, and the output layer of the Critic (i) is the estimated value function.

The algorithm was implemented in python programming language using PyTorch [27] and was based on the CleanRL [28] implementation of PPO for discrete action spaces. The default algorithm hyperparameters provided by the library were used.

3.5 Experimental Results

This section addresses all experimental results obtained. First, the case studies are presented, followed by the results of the reward function study and the PVT-inclusion approaches.

Fig. 3.6 Representation of the model used in this work: **a** Shared input layer, **b** First hidden layer of the actor network, **c** Second hidden layer of the actor network, **d** Third hidden layer of the actor network, **e** Output layer of the actor network **f** First hidden layer of the critic network, **g** Second hidden layer of the critic network, **h** Third hidden layer of the critic network, **i** Output layer of the critic network

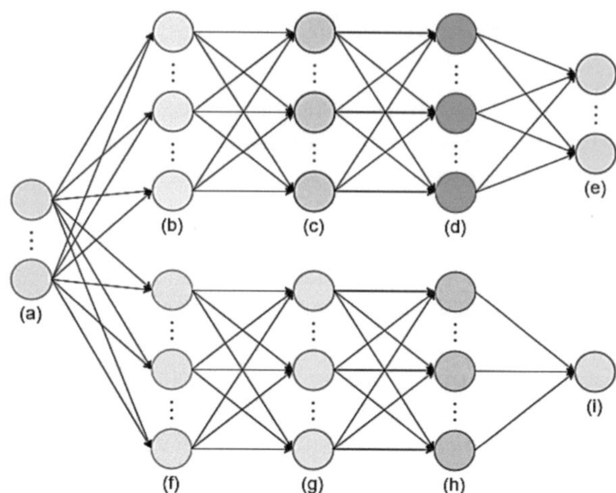

3.5.1 Case Studies

To conduct this extensive study, two different analog circuit topologies were selected. First, an amplifier using voltage combiners for gain enhancement (VCOTA) from [29], and, second, a folded cascade amplifier (FCA) used in [30], are considered. The VCOTA circuit schematic is illustrated in Fig. 3.7, and it uses PTM 130 nm technology with 8 PVT conditions, being the process corners {SS, FF, SF, FS}, supply corners {3.0, 3.6 V}, and temperature corners {0, 55 °C}. The FCA circuit schematic is illustrated in Fig. 3.8. Similarly, it uses PTM 130 nm technology with 8 PVT conditions, being the process corners {SS, FF, SF, FS}, supply corners {3.0, 3.6 V}, and temperature corners {0, 75 °C}. The NGSPICE simulator was employed to simulate/optimize these circuits, and the nominal simulation conditions are process {TT}, supply {3.3V} and temperature {25 °C}.

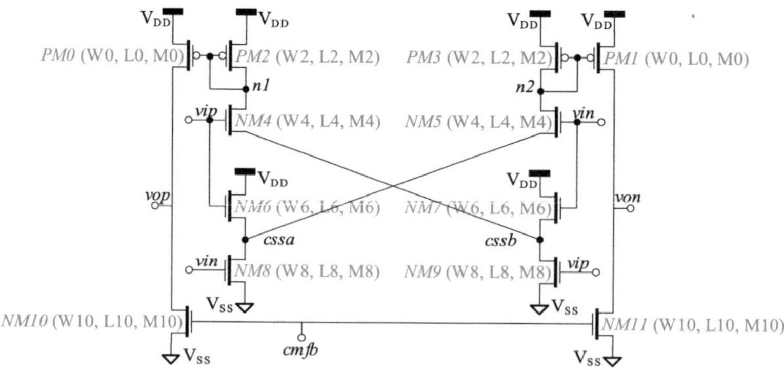

Fig. 3.7 Circuit schematic of the VCOTA

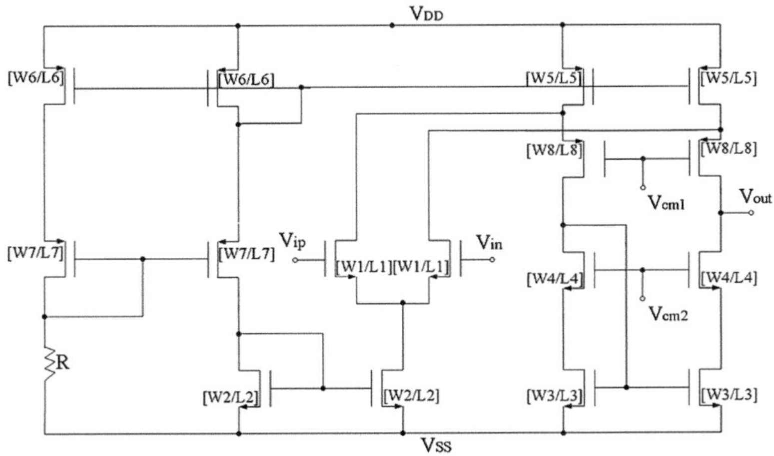

Fig. 3.8 Circuit schematic of the FCA. (Adapted from [30])

Table 3.1 presents the VCOTA ranges for the 18 design variables and Table 3.2 presents the performance figures of interest with the respective target goals. While Table 3.3 presents the FCA variables with respective ranges and Table 3.4 presents the performance figures of interest with the respective target goals. The custom FoM proposed for both circuits is:

$$FoM = \frac{GBW \times C_{load}}{Idd} \left[\frac{MHz \times pF}{mA} \right] \tag{3.11}$$

where GBW is the gain-bandwidth product, C_{load} is the load capacity, and Idd is the current consumption.

Table 3.1 VCOTA variables and ranges

Variable (unit)	Start	End	Step
Transistors width (μm)	1	100	1
transistor length (μm)	0.34	10.04	0.1
Transistor multiplicity	1	10	1

Table 3.2 VCOTA performance figures and target goals

ID	Unit	Description	Target goal
GBW	MHz	Gain-bandwidth product	≥ 1
GDC	dB	Low-frequency gain	≥ 40
Idd	μA	Current consumption	≤ 700
PM	Degree	Phase margin	$\geq 45°$ and $\leq 90°$

Table 3.3 FCA variables and ranges

Variable (unit)	Start	End	Step
Transistor width (μm)	1	100	1
Transistor length (μm)	0.34	10.04	0.1
Transistor multiplicity	1	10	1

Table 3.4 FCA performance figures and target goals

ID	Unit	Description	Target goal
GBW	MHz	Gain-bandwidth product	≥ 5
GDC	dB	Low-frequency gain	≥ 40
IDD	mA	Current consumption	≤ 1
PM	Degree	Phase margin	$\geq 45°$ and $\leq 90°$

3.5.2 Reward Functions

Starting with the VCOTA circuit, Table 3.5 presents the average results obtained over ten runs for the number of steps, number of runs that took less than 500,000 steps to converge, and training time, all with the same targets.

Between these five reward functions, the AutoCkt reward function comes as the best performing reward function, using the least number of steps, the least training time and successfully completing the ten runs. The PVTSizing reward function comes in close second, taking slightly more iterations, which translates to a slightly higher training time, but still manages to successfully complete all ten runs. These two results show the effectiveness of reward functions that include all target specifications and provide a smooth gradient that encourages the agent to move towards satisfying all specifications. Moving to the third best reward function, the steps reward succeeded in eight runs and achieved significantly worse results, almost $3\times$ the number of steps and training time of the AutoCkt reward function. This result proves what was expected. Even though the agent obtains some information about how it is performing in terms of failed specifications, this reward function does not help guiding the agent to a solution. Looking at the binary reward function, this reward provided the least information and, as expected, has the worst performance of all, not managing to

Table 3.5 VCOTA results for each reward function for 10 runs

Reward function	#Steps	#runs < 500 k steps	Training time (min)
Binary	–	0/10	–
Steps	106,373	8/10	60.30
AutoCkt	**35,867**	**10/10**	**20.62**
PVTSizing	36,777	**10/10**	22.146
FoM-based	348,024	3/10	193.56

complete a single run under the maximum steps allowed. This shows off the importance of providing meaningful information through the reward function. Finally, the custom FoM-based reward function did not have a significant performance either, taking $9.7\times$ the number of steps and training time of the AutoCkt reward function and not being able to finish all runs under the step limit. Table 3.6 presents the best solutions found in one of the complete runs for each of the reward functions. It is possible to observe that the AutoCkt reward function and the PVTsizing reward function achieved very similar results, with just a couple deviations. Interestingly, the steps reward function managed to get the best performing GDC, and the FoM-based reward function managed to achieve the highest GBW and the second lowest Idd. These two reward functions lack in some way a bias towards a certain specification. The steps reward function does not guide the agent to achieve some local or global minimum, and FoM-based only considers the GBW and Idd to reward the agent. This lack of guidance proved to worsen the training steps/training time, as seen previously, but seems to, in return, allow the agent to find better performing solutions in some specifications.

Figure 3.9 represents the running average reward at each step during training for each of the proposed reward functions. These values are not directly comparable, as each reward function has a different range of values, but it can provide insights on how the training is advancing. It is possible to observe that every reward function besides the FoM-based has an upwards tendency, clearly indicating that as training advances, the agent is improving. The FoM-based shows no upwards tendency, going up and down frequently, except at the end, where there still is a significant variation. This shows that the FoM computation does not clearly correlate with the agent achieving better results.

Additionally, Fig. 3.10 shows the evolution of each specification over a run using the trained nominal agent for each reward function on the VCOTA circuit. It is possible to see that the AutoCkt and the PVTSizing reward function show a very similar evolution of each specification, indicating that both reached a similar policy, as the actions taken by each at each step lead to similar behavior on the circuit. However, both the steps and the FoM-based reward functions exceeded the maximum steps per episode, taking the steps reward function a second episode to achieve the target goals and the FoM-based reward function three episodes to achieve the same target goals. This shows that, even though the agent reached the desired target goal, it did not achieve the same action proficiency as the best performing reward functions.

Table 3.6 VCOTA best specifications from each reward function

Reward function	GDC (dB)	GBW (MHz)	Idd (μA)	PM (°)
Binary	–	–	–	–
Steps	53.1	1.66	107	58.9
AutoCkt	43.9	1.99	62	48.7
PVTSizing	40.4	1.52	93	50.4
FoM-based	42.1	15.51	88	52.7

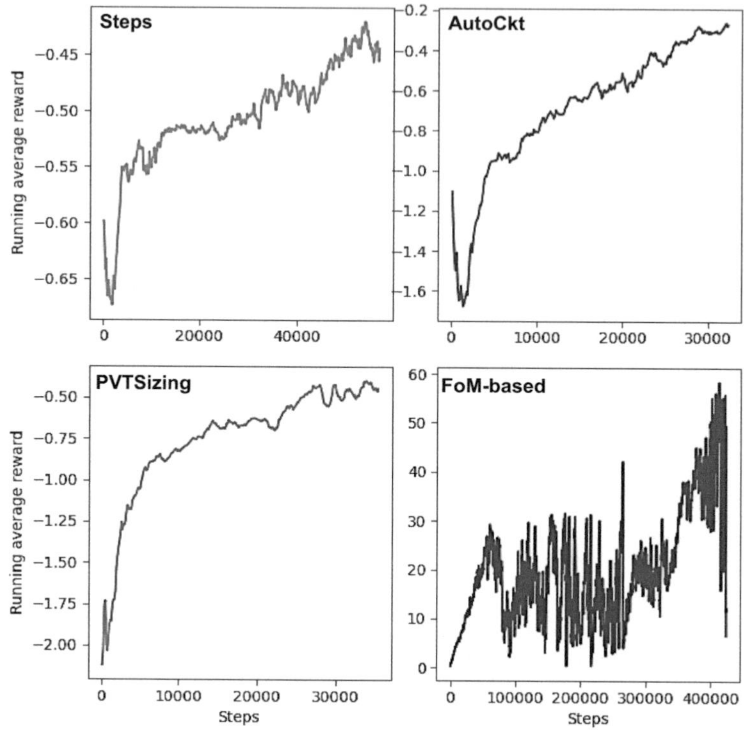

Fig. 3.9 Running average reward at each step during training for all reward functions for the VCOTA

Moving to the FCA circuit results, presented in Table 3.7, it is possible to observe that the overall conclusions are the same as the VCOTA results.

The AutoCkt reward function and the PVTSizing reward function remain the only ones that successfully completed the ten runs and achieved the best results, even though the gap between the two is slightly larger in this case study. The steps reward function continues to be the third best, the binary reward function continues to not be able to find a solution under the steps limit, and the FoM-based seems to have improved on this circuit, finishing now in seven out of the ten runs. This result could be due to a less demanding target objective, enabling more possible solutions to exist. However, it still takes significantly more training simulations and training time than the AutoCkt reward function, 5.13× and 5.36× , respectively. Table 3.8 presents the best solutions found in one of the complete runs for each of the reward functions.

It is possible to see that, for this circuit, the PVTSizing reward function and the AutoCkt reward function have a large difference in the GDC value, while everything else is similar. This could be just the randomness of RL or could be a sign that a lower constant positive reward does in fact help with exploration, leading to better solutions. For both circuits, it is evident that the lower constant reward led to a

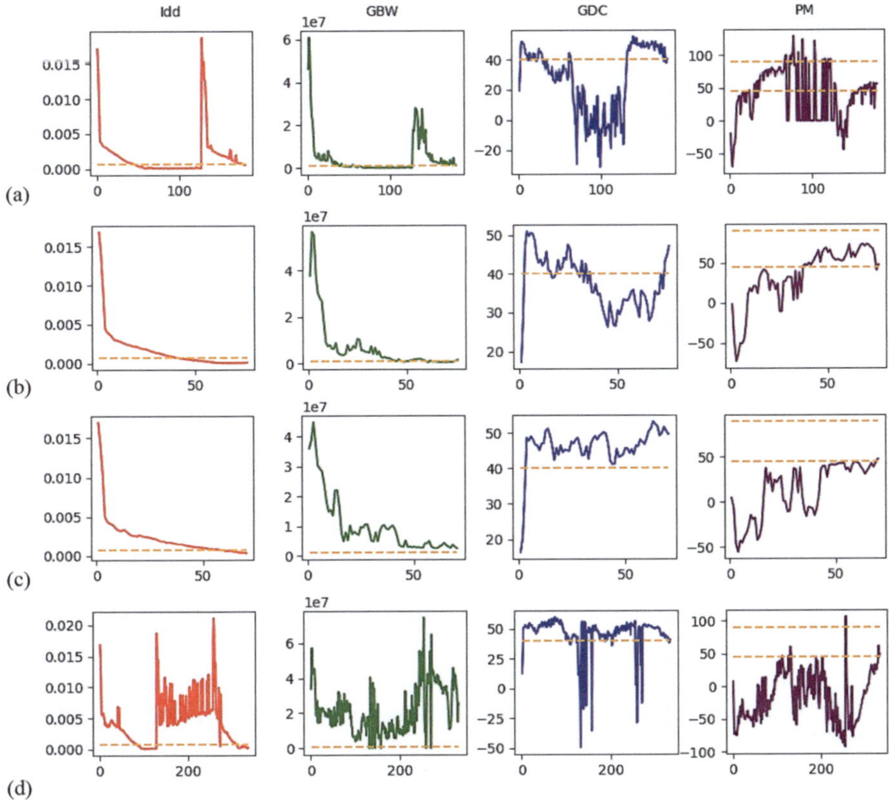

Fig. 3.10 Sample trajectory of the trained nominal agent to reach the target design specification on the VCOTA using different reward functions: **a** Steps; **b** AutoCkt; **c** PVT Sizing; and **d** FoM-based. The orange dashed lines correspond to the targets established

Table 3.7 FCA results for each reward functions for 10 runs

Reward function	#Steps	#runs < 500 k steps	Training time (min)
Binary	–	0/10	–
Steps	130,417	9/10	120.91
AutoCkt	**55,905**	**10/10**	**51.34**
PVTSizing	65,259	**10/10**	63.61
FoM-based	287,044	7/10	275.42

higher number of training samples/training time, but more research would be needed to confirm that it does indeed lead to better solutions. Moving to the steps reward function, in this circuit it did not excel in any specification, which could also be due to the randomness of RL. However, the FoM-based proved to achieve the highest GDC and GBW, being significantly better than the other reward functions, while

Table 3.8 FCA best specifications from each reward function

Reward function	GDC (dB)	GBW (MHz)	Idd (μA)	PM (°)
Binary	–	–	–	–
Steps	52.9	14.41	305	46.3
AutoCkt	58.6	18.74	571	65.7
PVTSizing	69.3	21.04	425	86.3
FoM-based	85.4	68.12	638	59.4

having the poorest Idd. This, once again, proves that, by including only the GBW and the Idd in the reward function, the agent tends to focus more on the trade-off of these specifications than equally balance all specifications. Figure 3.11 represents the running average reward at each step during training for each of the proposed reward functions. These values are not directly comparable, as each reward function has a different range of values, but it can give insights to how the training is advancing. Just like in the previous circuit, it is possible to observe that every reward function besides the FoM-based has an upwards tendency, further proving that the FoM computation does not clearly correlate with the agent achieving better results and, as such, should not be used as a reward function. Additionally, Fig. 3.12 shows the evolution of each specification over a run using the trained nominal agent for each reward function on the FCA circuit. It is possible to see that, for this circuit, all reward functions took one episode, proving that a good policy was obtained. The AutoCkt and the PVTSizing reward function continue to show a very similar evolution of each specification, and the Steps reward function takes the least number of steps for this circuit, hinting that this run achieved a faster converging policy.

3.5.3 *PVT Corner-Inclusive Approaches*

Based on the previous results, the PVTSizing reward function was selected. Even though the results were slightly worse than the AutoCkt reward function, Figs. 3.9 and 3.10 reveal that the reward evolution overtime was more stable and with less variance in the PVTSizing reward function than the AutoCkt reward function.

Starting with the VCOTA circuit, Table 3.9 presents the average results for number of steps, number of simulations and training time for each approach, all with the same targets, over ten different runs.

Between these five approaches, the brute force approach comes as the worst of all, in terms of number of simulations and time taken, which was expected. The reduced corner set and the worst-case approach have similar results, in number of simulations and time taken, but the number of steps is slightly different. The progressive overload comes in second and the scheduler comes as the best approach in terms of number of simulations and time taken, improving the brute force approach number of simulations by 4.7×. It is also curious to observe that the number of steps between the brute

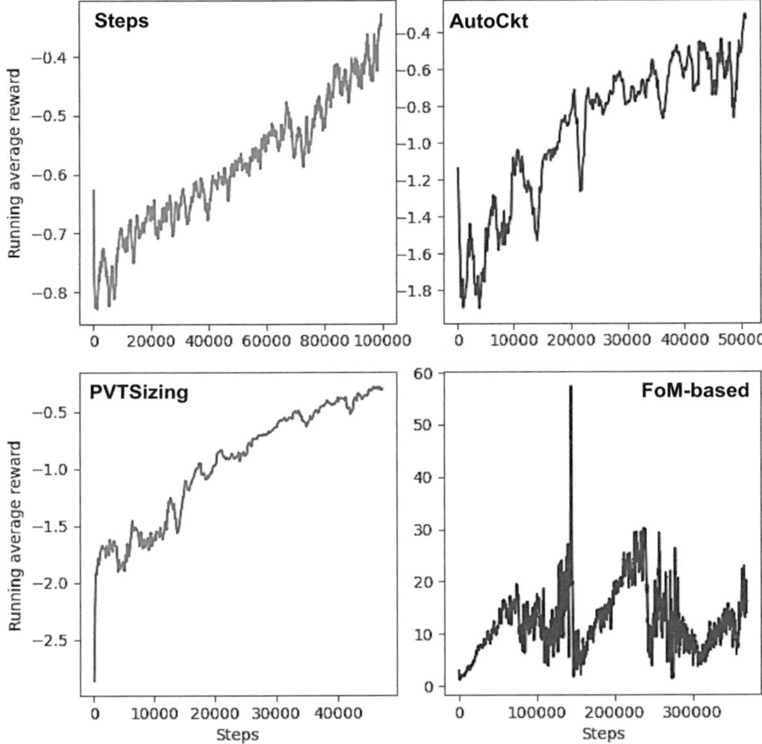

Fig. 3.11 Running average reward at each step during training for all reward functions for the FCA

force and the scheduler approach is the opposite. The difference in number of steps across all approaches could mean that, by giving the agent more information about the environment, the actions it takes are better, thus needing less steps. But, since the simulation cost of each step is so high, approaches that have less information at each step end up compensating by having more chances to try other candidate solutions. The low number of steps in the worst-case approach also suggests the importance of meaningful information. Table 3.10 presents the specifications obtained for the nominal corner for all approaches in this run.

It is possible to see that all approaches managed to reach similar specifications, even improving the starting specifications to accommodate the PVT corner variations. These results contribute to proving that the training was equivalent in every approach. Further analysis of the collected data also revealed that in the reduced corner set approach, where the full set was reduced to a three corner set, across all runs, the K-means algorithm always classified the voltage {3.6 V} and temperature {0 °C} corners as singular clusters and chose either the process {FS} or the temperature {55 °C} corner as the third element. In the worst-case approach, the process {FS} and the voltage {3.6 V} corners were always chosen and either of the temperature corners was chosen, depending on the run. This reveals that, for this circuit, this set of four

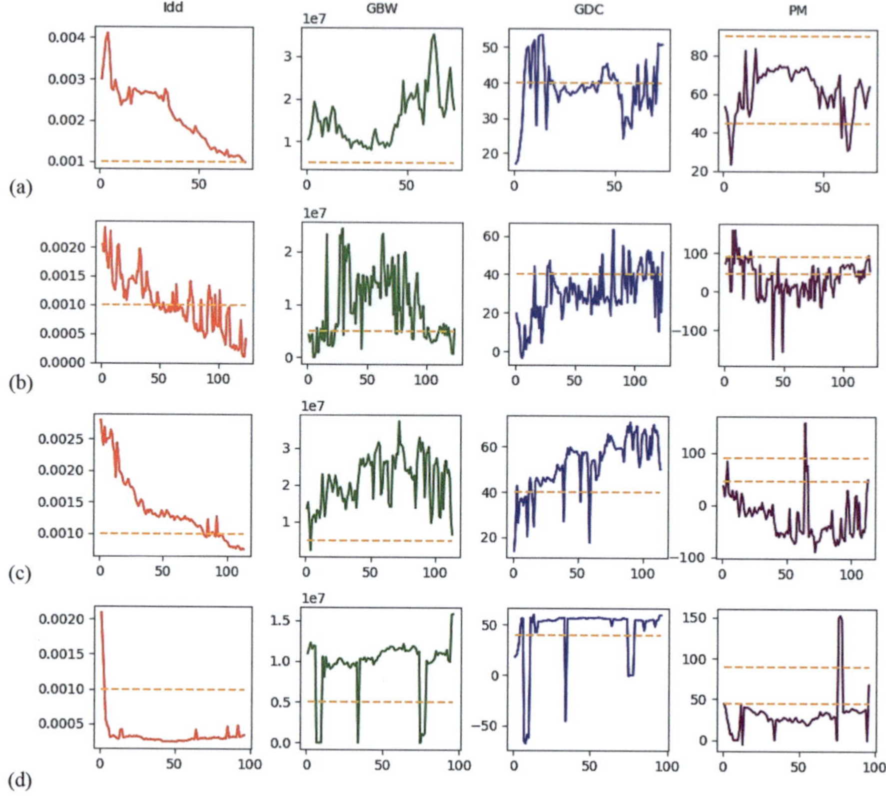

Fig. 3.12 Sample trajectory of the trained brute force agent to reach the target design specification on the FCA using different reward functions: **a** Steps; **b** AutoCkt; **c** PVT Sizing; and **d** FoM-based. The orange dashed lines correspond to the targets established

Table 3.9 VCOTA results for each approach for 10 runs

Approach	#Steps	#Simulations	Training time (min)
Brute force	43,514	391,626	200.13
Progressive overload	47,492	126,636	68.28
Reduced corner set	53,339	160,559	85.54
Scheduler	76,796	**83,943**	**49.92**
Worst case	**41,769**	171,117	90.06

corners is well representative of all possible variations, thus encouraging the use of approaches as the ones described in this work to reduce the number of simulations done when incorporating PVT corner conditions on RL-based agents. Figure 3.13 presents the running average reward at each step of training for all approaches. This data corresponds to one of the ten runs.

Table 3.10 VCOTA best nominal corner specifications from each approach

Approach	GDC (dB)	GBW (MHz)	Idd (µA)	PM (°)
Starting point	41.3	3.80	275	51.6
Brute force	54.5	7.84	309	55.8
Progressive overload	49.4	6.22	205	57.6
Reduced corner set	48.5	8.17	219	52.2
Scheduler	51.2	5.97	267	52.7
Worst case	52.0	4.75	202	55.4

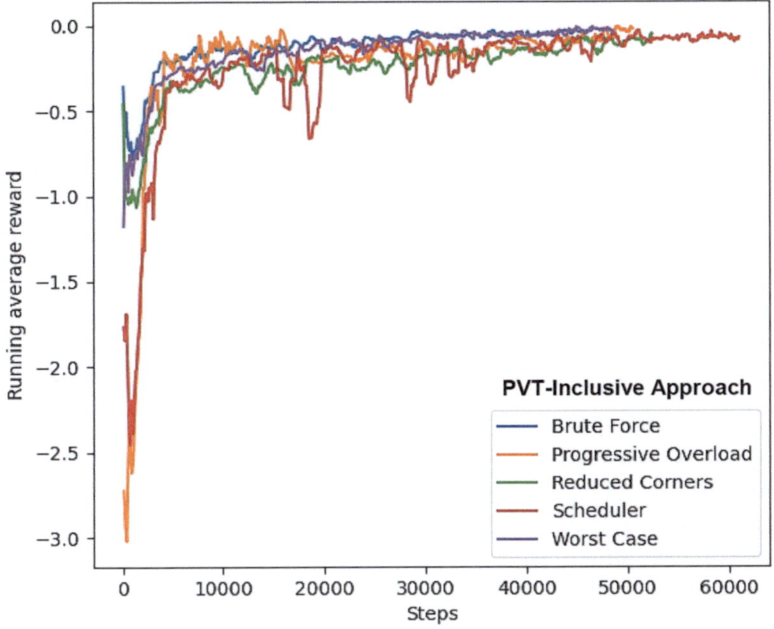

Fig. 3.13 VCOTA: running average reward at each step during training for all approaches. This plot data corresponds to one of the ten runs

It is possible to observe that approaches with less information start worse, as the agent is still determining how to use the nominal knowledge to this condition not seen before, but as training evolves, all methods end up reaching similar levels of proficiency. Figure 3.14 shows the number of simulations done per step for all approaches. This data was obtained by running the trained agent of each approach until it managed to successfully size the circuit.

Additionally, Figs. 3.15, 3.16, 3.17, 3.18 and 3.19 present the results of running each approach trained agent from the starting point given by the nominal agent to a solution that accommodates all corners. In Fig. 3.15, Idd remains consistently below the target across all corners. However, the agent struggles to balance the remaining

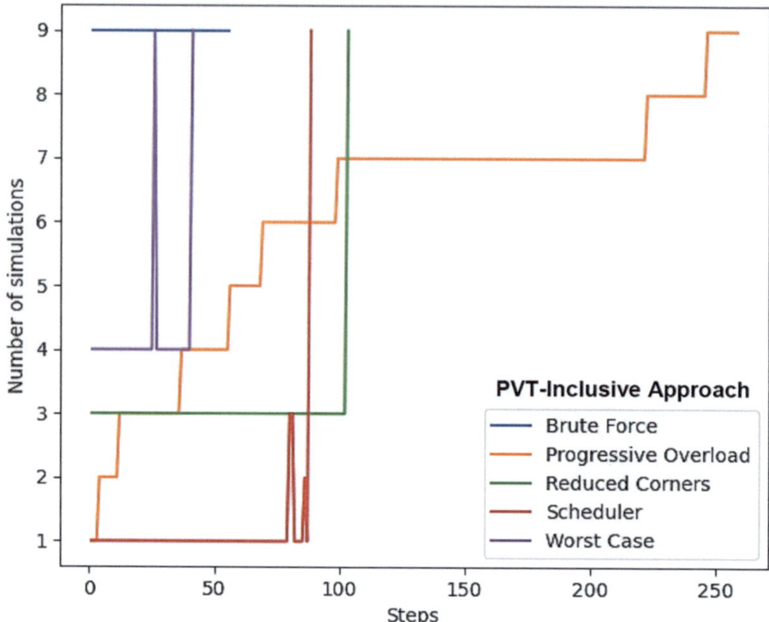

Fig. 3.14 VCOTA: number of simulations done at each step from starting point to successful sizing for all approaches. This plot data was obtained using the trained model of each approach

three specifications. Particularly, it has difficulty meeting both the GBW and the GDC in the voltage {3.6 V} corner, and the PM in the temperature {0 °C} corner, which are two of the corners usually selected by the reduced corner set approach and the worst-case approach. Interestingly, for each specification, the curves across corners are similar in shape, differing primarily by a slight upward or downward offset, which indicates that the corners are variations of the nominal values. Figure 3.16 illustrates the progressive corner inclusion approach. As new corners are added, the agent must iteratively readjust the circuit to meet target values. Despite this adaptation process, the similarity among the curves remains evident. In Fig. 3.17, the reduced corner set optimization process is shown. The agent first balances performance across the three selected corners before running a full simulation. While Idd, GBW, and GDC are met with minimal difficulty, the PM target remains challenging to achieve. Figure 3.18 shows the behavior of the scheduler method, which dedicates most of its effort to satisfying the voltage {3.6 V} corner. Once a solution is found for this corner, the agent quickly converges to meet the targets in the remaining corners. Finally, in Fig. 3.19, it is possible to see that the agent begins by balancing the four worst-case corners and generates an initial candidate solution, which fails the PM requirement. This leads to further refinement until a final solution meets all specification targets across all corners.

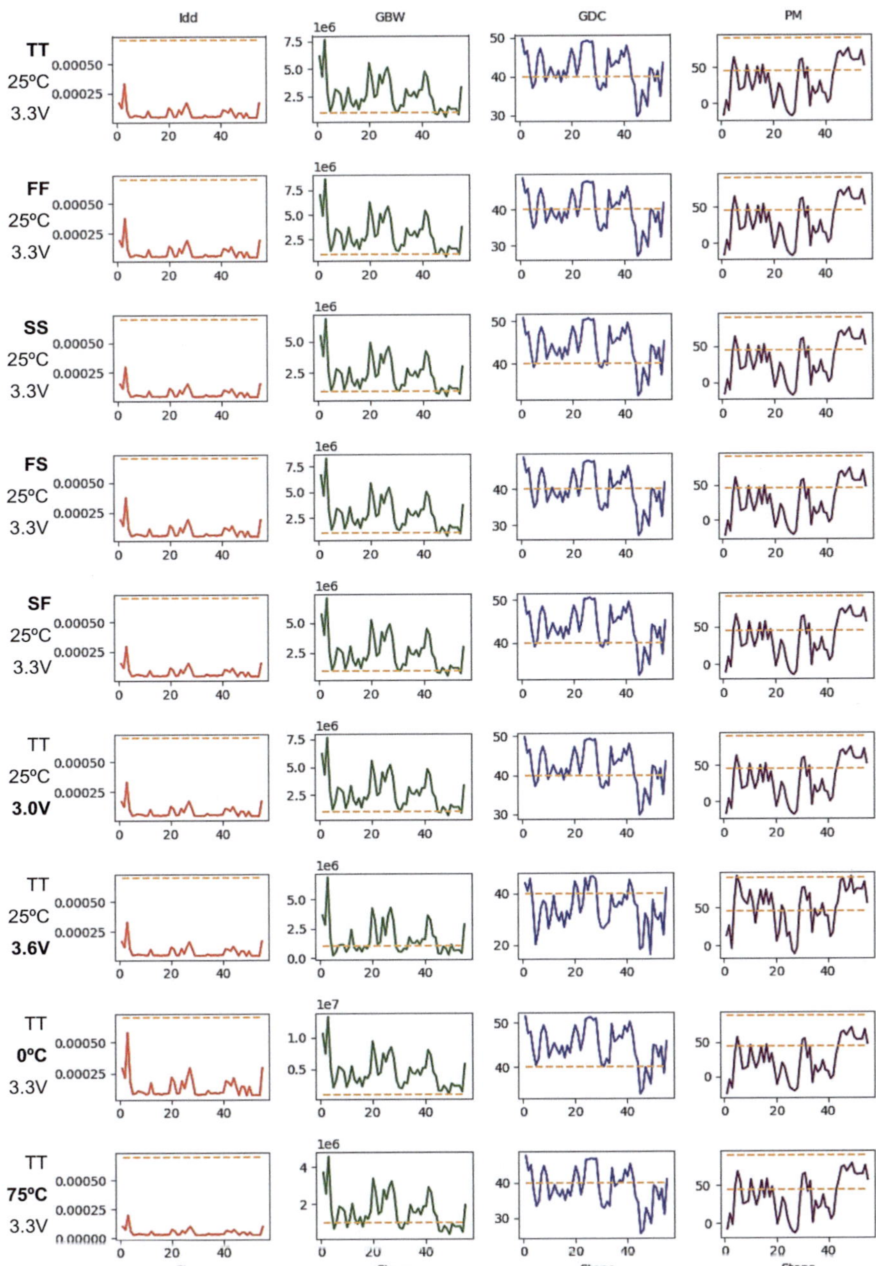

Fig. 3.15 Sample trajectory of the trained Brute Force agent to reach the target design specification on the VCOTA. The orange dashed lines correspond to the targets established

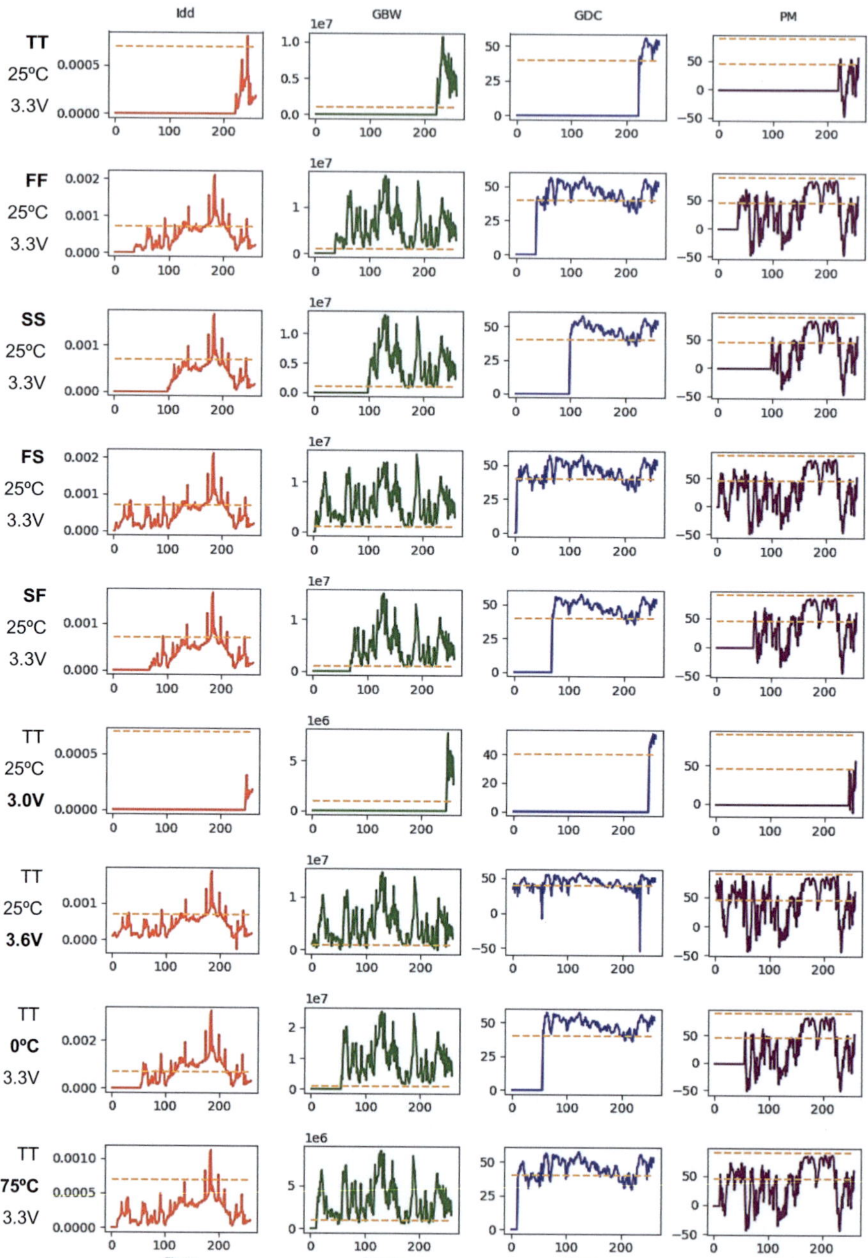

Fig. 3.16 Sample trajectory of the trained Progressive Overload agent to reach the target design specification on the VCOTA. The orange dashed lines correspond to the targets established

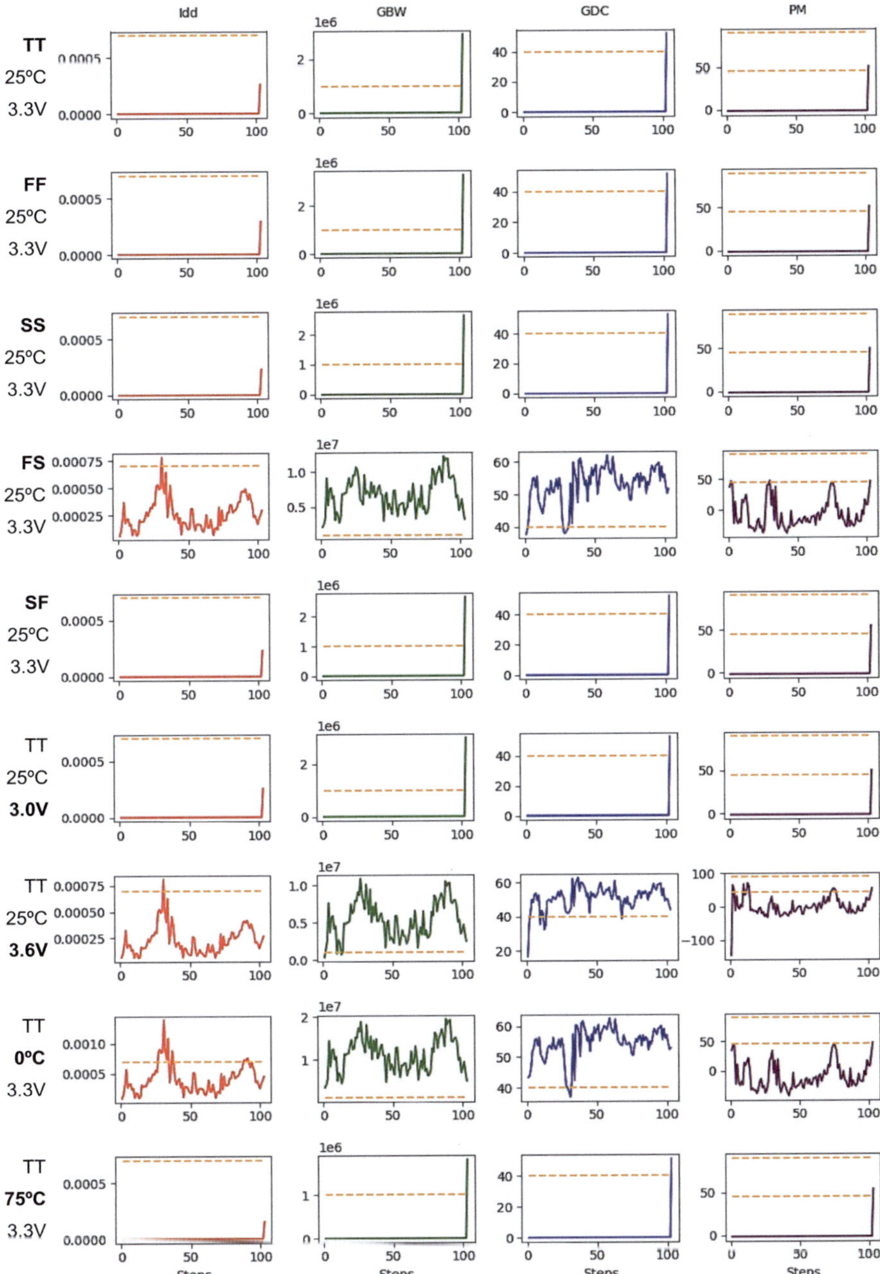

Fig. 3.17 Sample trajectory of the trained Reduced Corner Set agent to reach the target design specification on the VCOTA. The orange dashed lines correspond to the targets established

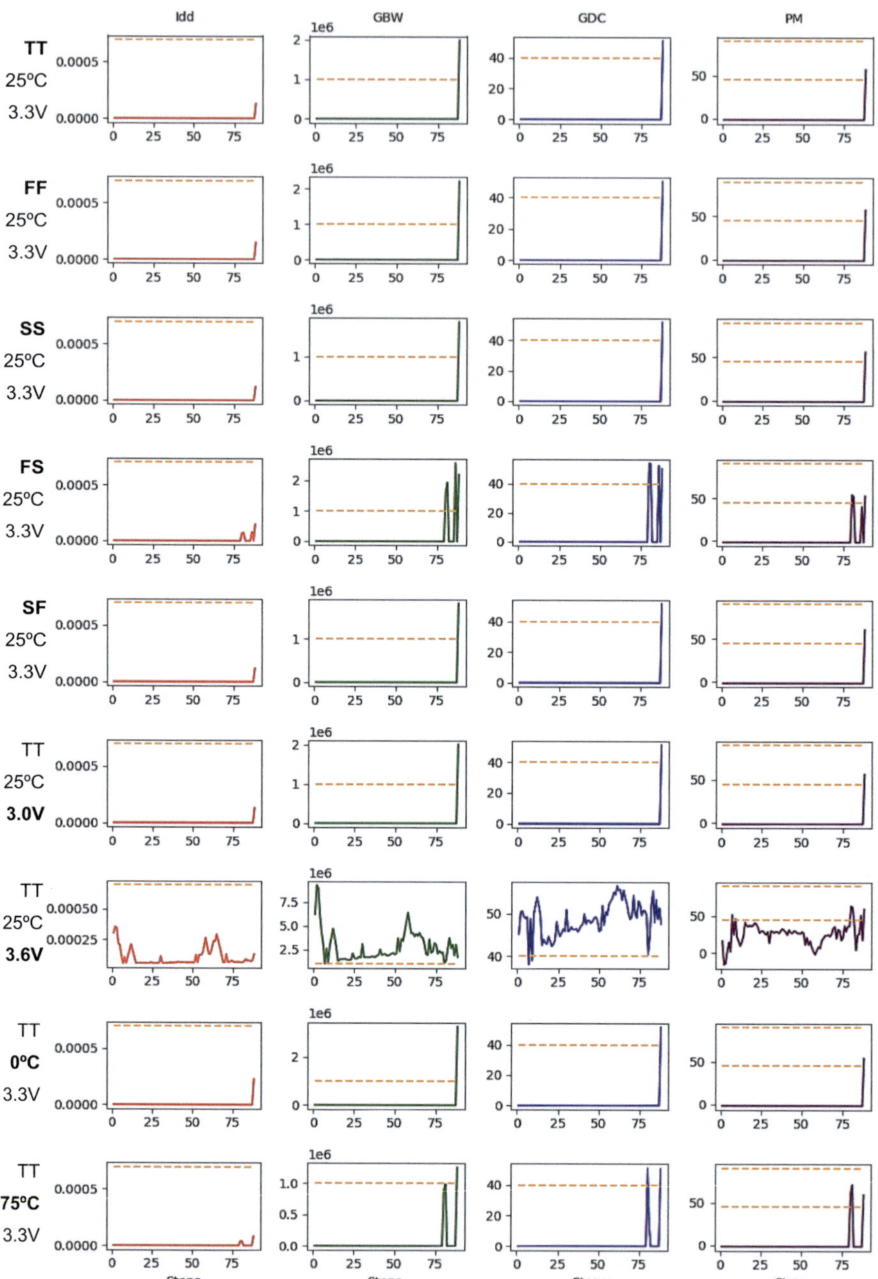

Fig. 3.18 Sample trajectory of the trained Scheduler agent to reach the target design specification on the VCOTA. The orange dashed lines correspond to the targets established

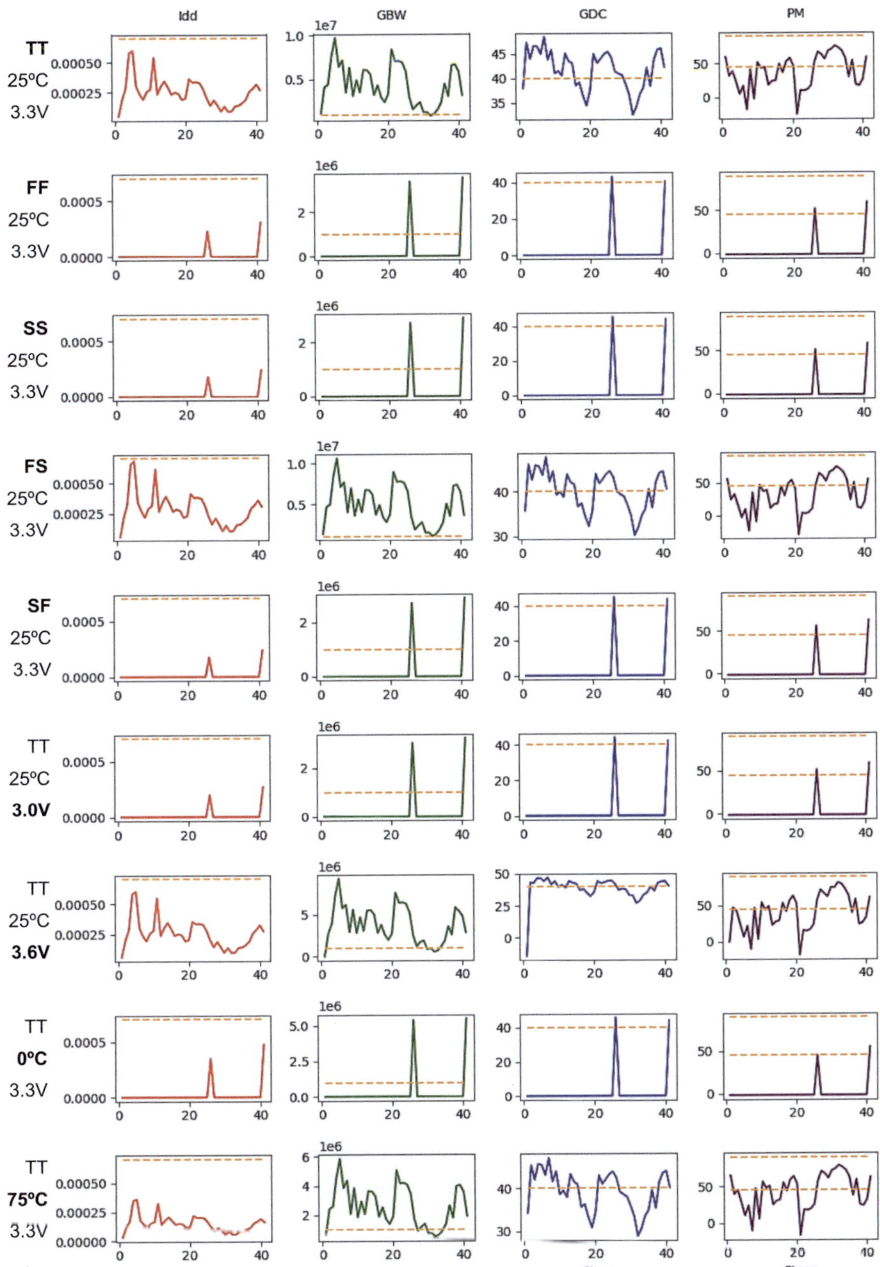

Fig. 3.19 Sample trajectory of the trained Worst-Case agent to reach the target design specification on the VCOTA. The orange dashed lines correspond to the targets established

Moving to the FCA circuit, Table 3.11 presents the average results for number of steps, number of simulations, and training time for each approach, all with the same targets, over ten different runs.

Just like the previous circuit, the brute force approach comes as the worst of all, in terms of number of simulations and time taken. Opposite to the previous circuit results, the progressive overload approach performs significantly worse, coming in fourth this time. The reduced corner set and the worst-case approach also show a bigger discrepancy in this circuit, in the number of simulations and time taken. The scheduler continues to be the best approach in terms of number of simulations and time taken, improving the brute force approach number of simulations by 3.2×. It is also curious to observe that the difference in number of steps across all approaches, apart from the worst case, still follows the behavior of the previous circuit, meaning that more information about the environment translates into better actions and less steps. Table 3.12 presents the specifications obtained for the nominal corner for all approaches in this run.

It is possible to see that all approaches managed to reach similar specifications, some having one specification a little above the others, but it still is possible to see an improvement of the starting specifications to accommodate the PVT corner variations. These results also contribute to proving that the training was equivalent in every approach. Conducting the same analysis of the collected data for this circuit revealed that in the reduced corner set approach, where the full set was reduced to a three corner set, across all runs, the K-means algorithm only selected the temperature {75 °C} corner as singular cluster in every run. For the second corner, the algorithm had a big preference for the process {FS} corner, although it did not choose it in every

Table 3.11 FCA results for each approach for 10 runs

Approach	#Steps	#Simulations	Training Time (Min)
Brute force	**39,345**	354,105	296.89
Progressive overload	51,402	250,443	219.79
Reduced corner set	43,067	136,473	120.81
Scheduler	63,910	**111,725**	**104.27**
Worst case	45,696	195,792	166.93

Table 3.12 FCA best nominal corner specifications from each approach

Approach	GDC (dB)	GBW (MHz)	Idd (μA)	PM (°)
Starting point	47.1	8.09	354	77.7
Brute force	54.1	12.86	355	56.2
Progressive overload	55.5	10.39	247	61.3
Reduced corner set	59.6	19.06	489	54.5
Scheduler	67.3	11.66	251	57.7
Worst case	68.8	14.05	300	52.6

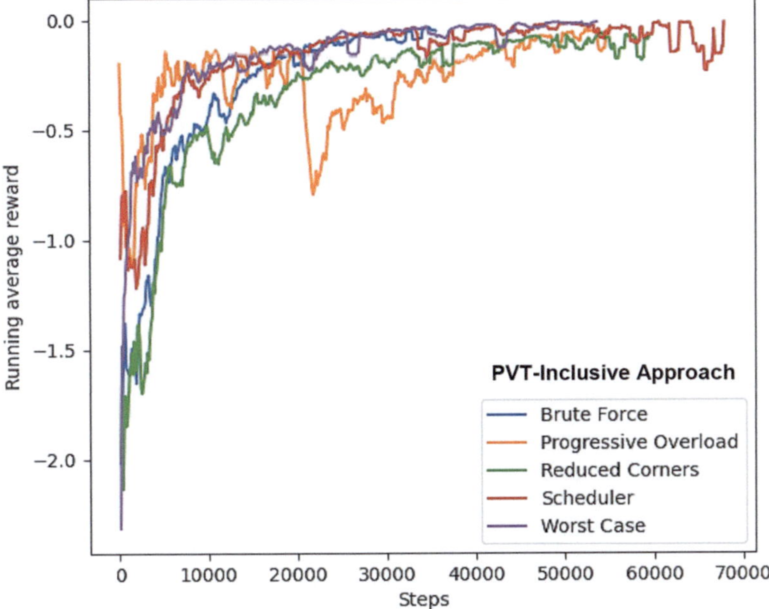

Fig. 3.20 FCA: running average reward at each step during training for all approaches. This plot data corresponds to one of the ten runs

run. The third corner varied a lot in choice. In the worst-case approach, there were no corners being chosen in every run, but there was a tendency to choose the process {SF} corner, the voltage {3.6 V} corner and the temperature {75 °C} corner. This reveals that, for this circuit, there is no clear set of corners that are representative of all possible variations, thus leading to the higher variations and difficulty found during training. Figure 3.20 presents the running average reward at each step of training for all approaches. This data corresponds to one of the ten runs.

This time it is possible to see the opposite of the previous circuit. Approaches with less information start better, but, in the end, all methods end up reaching similar levels of proficiency. For this circuit, the progressive overload approach struggles a lot more than on the previous circuit. Around step 20,000, the average reward per step has a significant drop in value, probably due to the introduction of a new corner. Figure 3.21 shows the number of simulations done per step for all approaches. This data was obtained by running the trained agent of each approach until it managed to successfully size the circuit. It is possible to see that this time, the scheduler and the worst case struggled a little more than the previous circuit, while other approaches had similar behavior.

Additionally, Figs. 3.22, 3.23, 3.24, 3.25 and 3.26 present the results of running each approach trained agent from the starting point given by the nominal agent to a solution that accommodates all corners. Starting with Fig. 3.22, BW remains consistently above the target across all corners. However, the agent struggled to meet

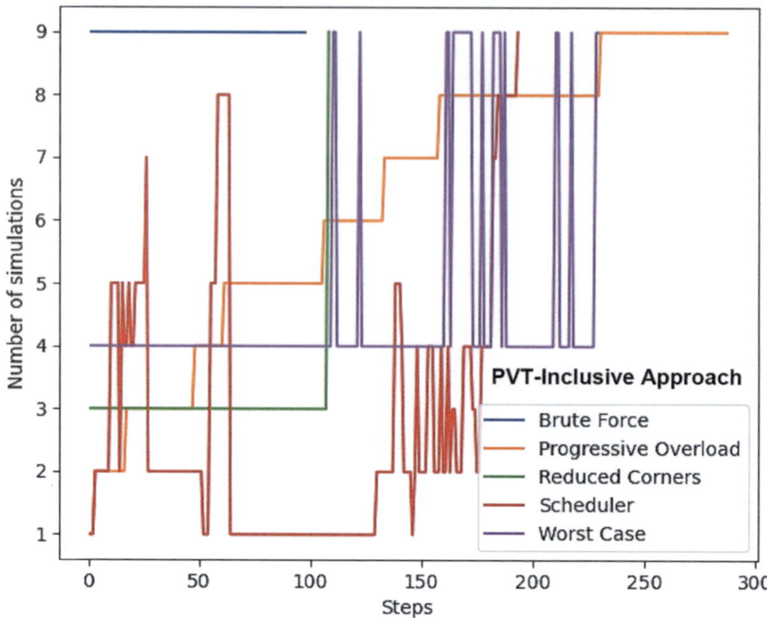

Fig. 3.21 FCA: number of simulations done at each step from starting point to successful sizing for all approaches. This plot data was obtained using the trained model of each approach

the Idd requirement in the temperature {0 °C} corner had difficulty achieving the PM target in a few corners. In Fig. 3.23, the agent begins with the temperature {0 °C} corner, again encountering the same Idd issue. Later, when the temperature {75 °C} corner is introduced, the agent also struggles to meet the PM target. Despite these challenges, both GBW and GDC were generally maintained above their target values. Figure 3.24 shows a different outcome. The agent had no significant issues with Idd. It's important to note that all runs begin from the same starting point, using the same nominal agent and seed. Nonetheless, the results differ in this case. While the agent continued to struggle with the PM in the temperature {75 °C} corner but, once the three selected corners met their target specifications, the remaining corners quickly followed. In Fig. 3.25, the scheduler approach shows a significant variation in the number of corners being simulated at each step. The agent initially avoided Idd issues, but as it refined other corners, an Idd violation reappeared. Resolving this required a considerable number of steps before a valid solution could be achieved. GBW and GDC generally stayed above their targets when simulated, but PM continued to pose difficulties. Finally, in Fig. 3.26, the agent experienced no issues with Idd. GBW and GDC again remained mostly above target, but PM remained a persistent challenge. In this run, the agent performed a large number of full-corner simulations before eventually arriving at a solution that satisfied all specifications.

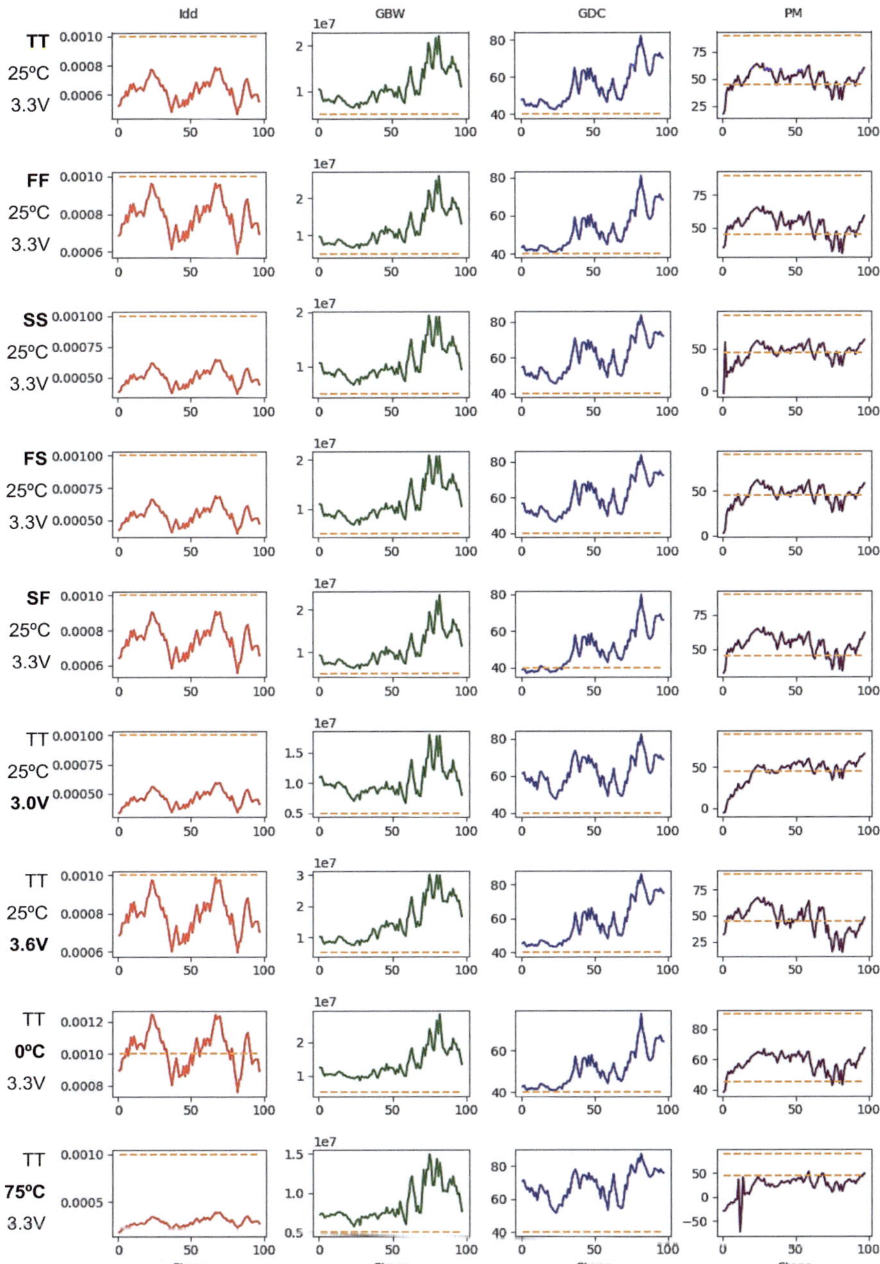

Fig. 3.22 Sample trajectory of the trained Brute Force agent to reach the target design specification on the FCA. The orange dashed lines correspond to the targets established

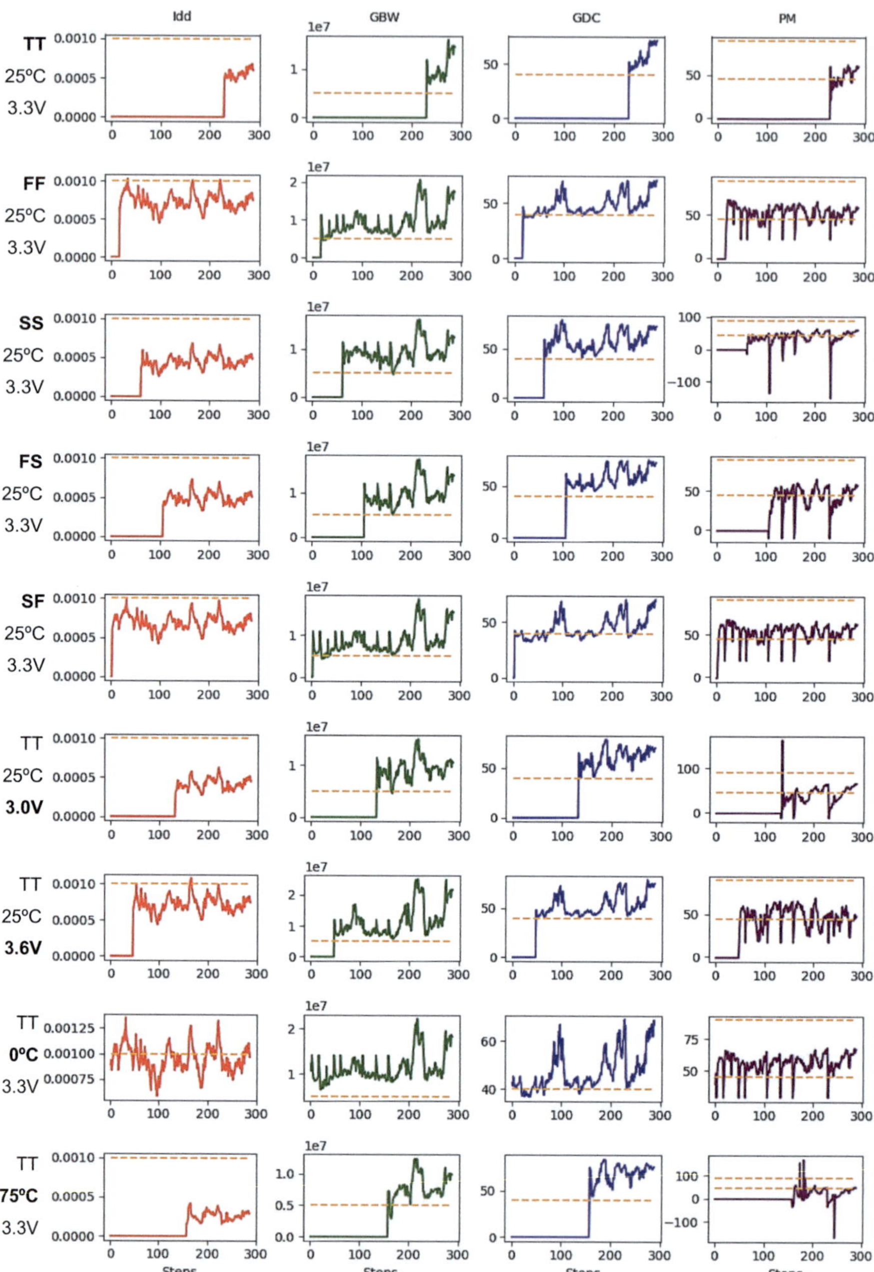

Fig. 3.23 Sample trajectory of the trained Progressive Overload agent to reach the target design specification on the FCA. The orange dashed lines correspond to the targets established

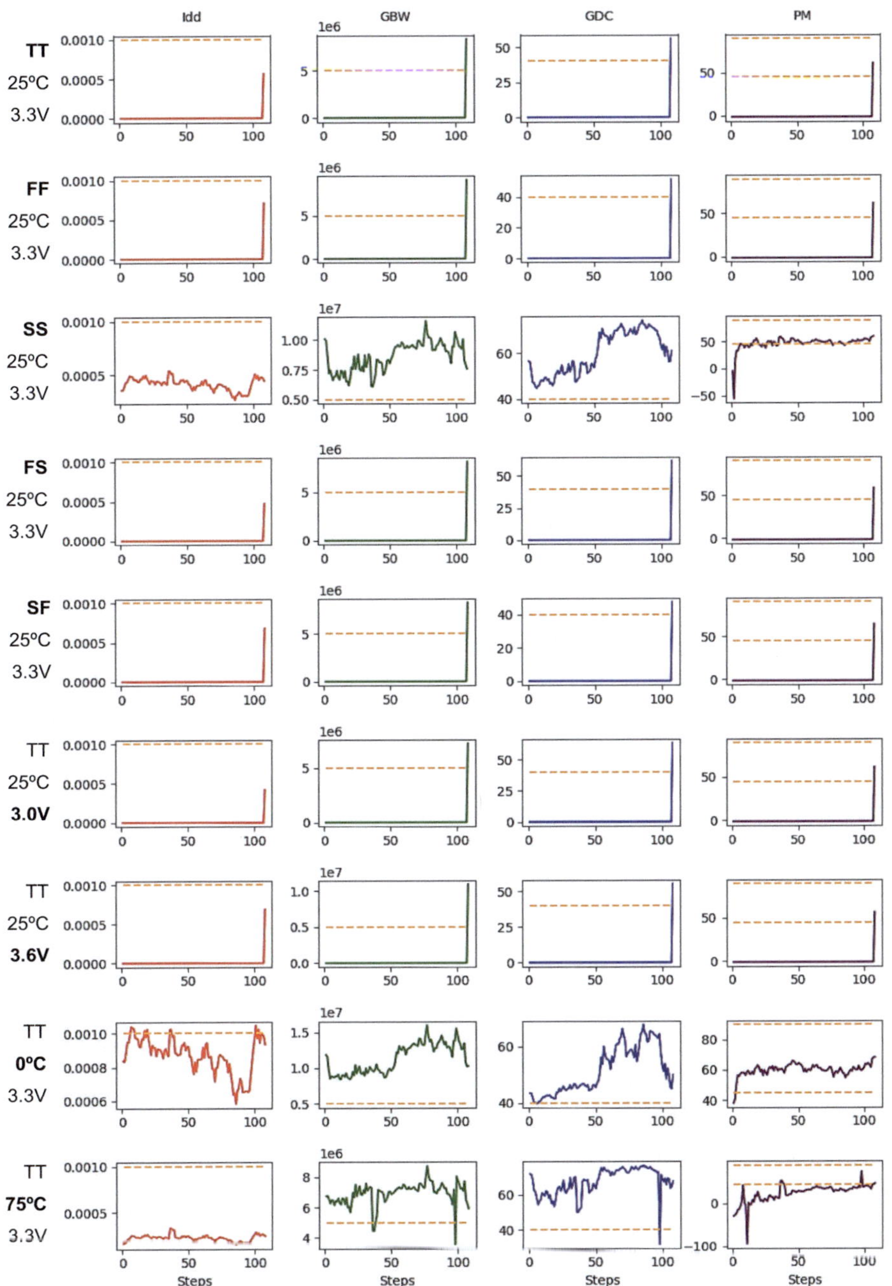

Fig. 3.24 Sample trajectory of the trained Reduced Corner Set agent to reach the target design specification on the FCA. The orange dashed lines correspond to the targets established

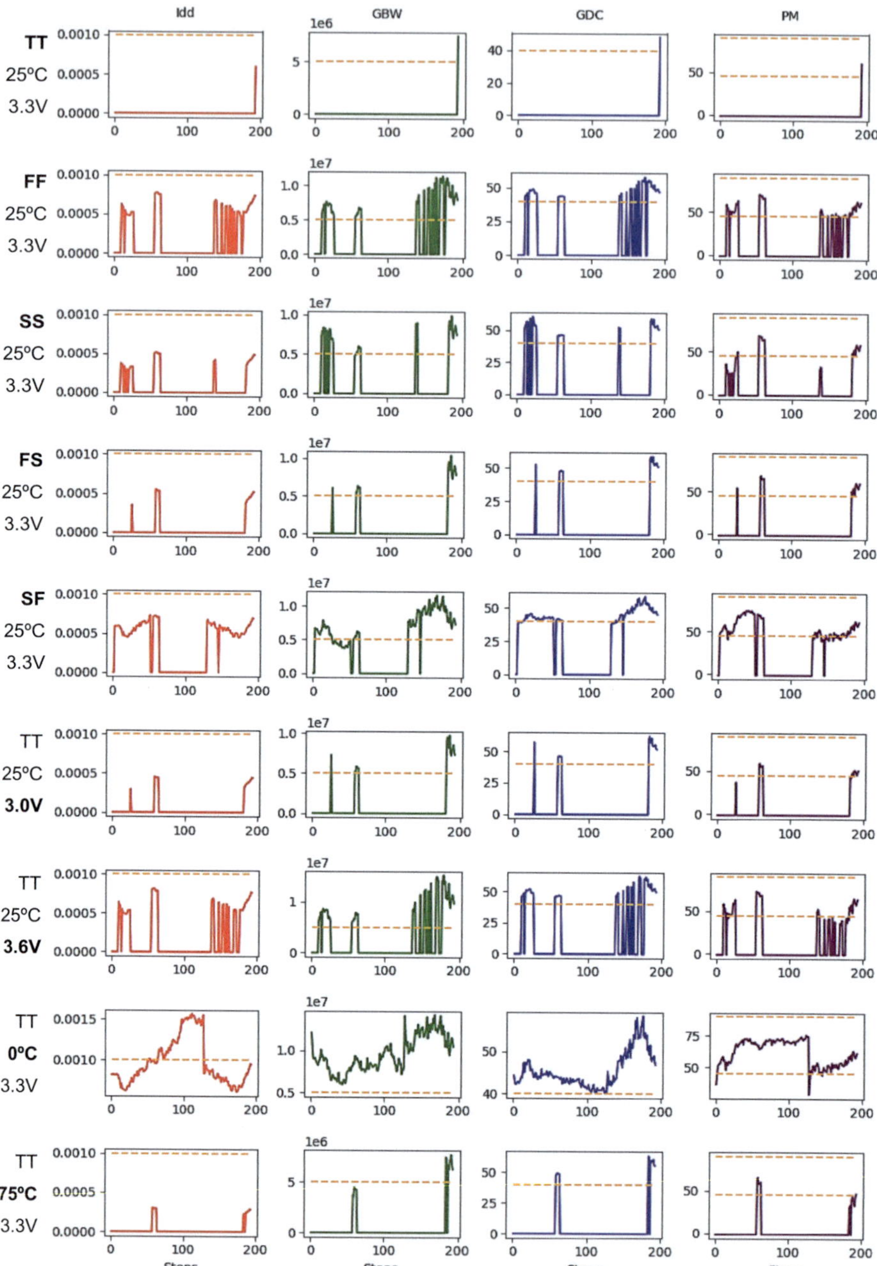

Fig. 3.25 Sample trajectory of the trained Scheduler agent to reach the target design specification on the FCA. The orange dashed lines correspond to the targets established

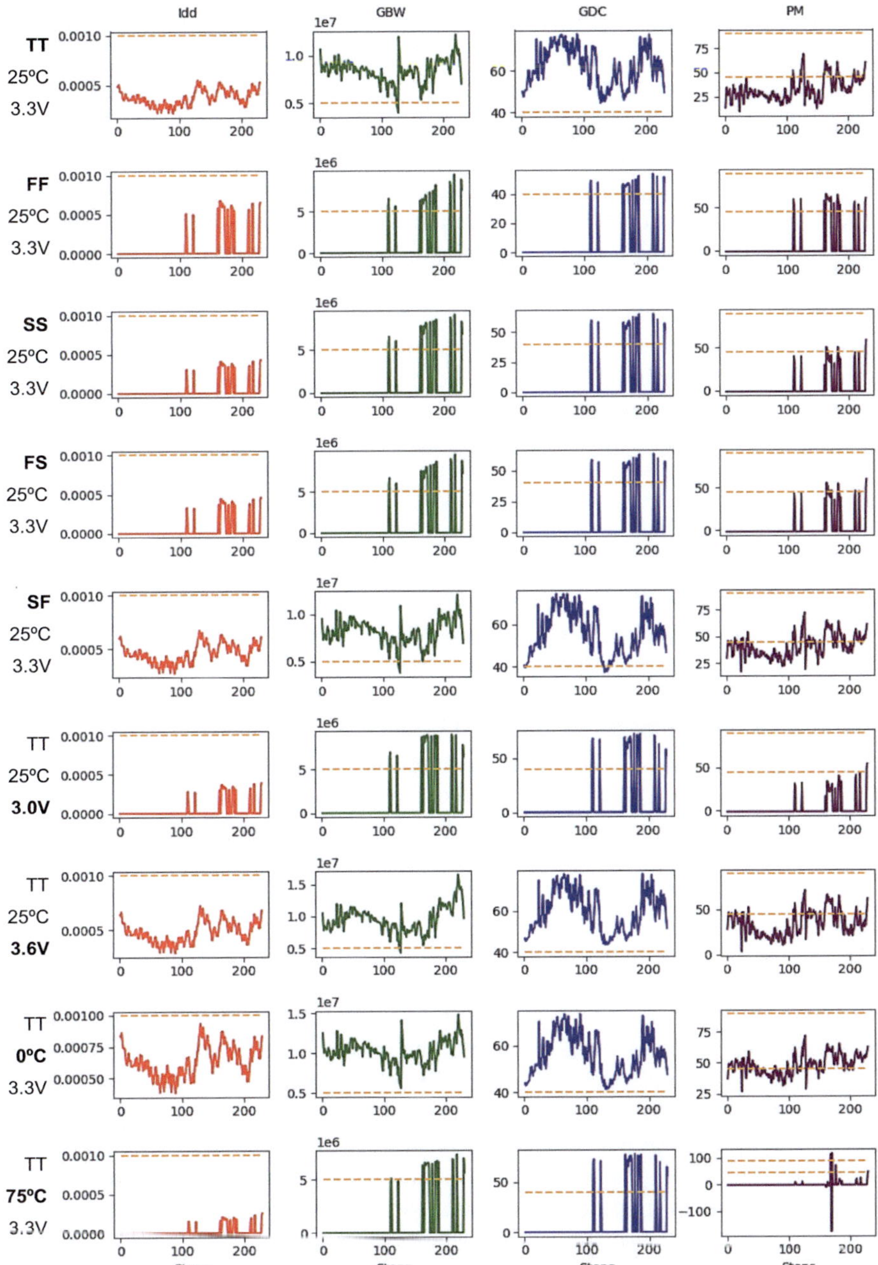

Fig. 3.26 Sample trajectory of the trained Worst-Case agent to reach the target design specification on the FCA. The orange dashed lines correspond to the targets established

3.6 Conclusions and Future Research Directions

In this Chapter, five different approaches for PVT corner incorporation were compared, i.e., brute force approach, progressive overload approach, reduced corner set, scheduler, and worst-case corner. Out of all, the scheduler approach demonstrated the most promising simulation efficiency, up to $3.7\times$ less simulations than the brute force approach, used in other works. Even though this approach takes more steps to reach proficiency, which translates to sampling a larger number of possible solutions and usually associated with longer convergence times, the number of simulations per step is always the minimum necessary, due to only simulating a corner if the previous reached the target specifications, making the number of steps for speed increase trade-off worth it. Other data collected also support that the agent is equally as capable as other approaches.

3.6.1 Conclusions

This study presented a RL agent robust to PVT conditions where five different reward functions and five different approaches for PVT corner incorporation were compared. Among the five reward functions, i.e., the binary reward, the steps reward, the reward function from AutoCkt, the reward function from PVTSizing, and the custom FoM-based reward, the AutoCkt and the PVTSizing reward functions showed superior training capabilities and simulation efficiency, having results up to $9.7\times$ better on some test cases. Although the AutoCkt reward function proved to achieve proficiency quicker than the PVTSizing reward function, and both proved to reach a similar policy, the PVTSizing reward function was chosen as the better one due to the greater stability and less variance seen overtime in the running average reward at each step during training.

Out of all the PVT-inclusive approaches, i.e., brute force approach, progressive overload approach, reduced corner set, scheduler, and worst-case corner, the scheduler approach demonstrated the most promising simulation efficiency, up to $4.7\times$ less simulations than the brute force approach, used in other works. Even though this approach takes more steps to reach proficiency, which translates to sampling a larger number of possible solutions and usually associated with longer convergence times, the number of simulations per step is always the minimum necessary, due to only simulating a corner if the previous reached the target specifications, making the number of steps for speed increase trade-off worth it. Other data collected also supports that the agent is equally as capable as other approaches.

3.6.2 *Future Work*

As seen in Chap. 2 of this book, there are several of different approaches to tackling the analog IC sizing problem, each employing its own technique. Some suggestions that could expand this work are outlined as follow: (1) exploring the usage of a different neural network structure. This could be the usage of more layers, a different number of neurons, or even a different structure, for example graph neural networks; (2) employment of a multi-agent approach is also appealing, but from the research conducted, no other work tackles this approach. A further study to compare the pros and cons is to be pursue. Multi-agent works also did not consider PVT variations, thus understanding the implications these bring in this context would be valuable; and (3) one common understanding that all runs among all approaches done in this work showed was the importance of a good starting point. The exploration of other methods to quickly achieve a satisfactory starting point would be of great value. Some of the prior works analyzed employed BO to achieve this end, but there might be other techniques that can achieve even better results.

References

1. Mina R, Jabbour C, Sakr G (2022) A review of machine learning techniques in analog integrated circuit design automation. Electron MDPI 11(3):435
2. Martins R, Lourenço N (2023) Analog integrated circuit routing techniques: an extensive review. IEEE Access 11:35965–35983
3. Wang C, Yang F, Zhu K (2024) AI-enabled layout automation for analog and RF IC: current status and future directions. In: Proceedings of the IEEE international symposium on radio-frequency integration technology
4. Maji S, Budak A, Poddar S, Pan D (2024) Toward end-to-end analog design automation with ML and data-driven approaches (invited paper). In: Proceedings of the 29th Asia and South Pacific design automation conference
5. Martins R (2025) A survey of machine and deep learning techniques in analog integrated circuit layout synthesis. Microelectron MDPI 1(1, 2)
6. Suissa A et al (2010) Empirical method based on neural networks for analog power modeling. IEEE Trans Comput Aided Des Integr Circ Syst 29(5):839–844
7. Wolfe G, Vemuri R (2003) Extraction and use of neural network models in automated synthesis of operational amplifiers. IEEE Trans Comput Aided Des Integr Circ Syst 22(2):198–212
8. Alpaydin G, Balkir S, Dundar G (2003) An evolutionary approach to automatic synthesis of high-performance analog integrated circuits. IEEE Trans Evol Comput 7(3):240–252
9. Liu H, Singhee A, Rutenbar R, Carley L (2002) Remembrance of circuits past: macromodeling by data mining in large analog design spaces. In: Proceedings 2002 design automation conference, pp 437–442
10. Azevedo F, Lourenço N, Martins R (2025) Comprehensive application of denoising diffusion probabilistic models towards the automation of analog integrated circuit sizing. Expert Syst Appl 290:128414
11. Eid P, Azevedo F, Lourenço N, Martins R (2025) Using denoising diffusion probabilistic models to solve the inverse sizing problem of analog integrated circuits. AEU—Int J Electron Commun 195:155767
12. Zhu K et al (2019) Genius route: a new analog routing paradigm using generative neural network guidance. In: Proceedings of the IEEE/ACM international conference on computer-aided design

13. Gusmão A, Horta N, Lourenço N, Martins R (2022) Scalable and order invariant analog integrated circuit placement with attention-based graph-to-sequence deep models. Expert Syst Appl 207:117954

14. Gusmão A, Póvoa R, Horta N, Lourenço N, Martins R (2022) DeepPlacer: a custom integrated OpAmp placement tool using deep models. Appl Soft Comput 115:108188

15. Gusmão A, Horta N, Lourenço N, Martins R (2021) Late breaking results: attention in Graph2Seq neural networks towards push-button analog IC placement. In: ACM/IEEE design automation conference

16. Andraud M, Stratigopoulos H, Simeu E (2016) One-shot non-intrusive calibration against process variations for analog/RF circuits. IEEE Trans Circ Syst I Regul Pap 63(11):2022–2035

17. Settaluri K, Liu Z, Khurana R, Mirhaj A, Jain R, Nikolic B (2022) Automated design of analog circuits using reinforcement learning. IEEE Trans Comput Aided Des Integr Circ Syst 41(9):2794–2807

18. Bao J et al (2024) Multiagent Based Reinforcement Learning (MA-RL): an automated designer for complex analog circuits. IEEE Trans Comput Aided Des Integr Circ Syst 43(12):4398–4411

19. Cao W, Gao J, Ma T, Ma R, Benosman M, Zhang X (2025) RoSE-Opt: robust and efficient analog circuit parameter optimization with knowledge-infused reinforcement learning. IEEE Trans Comput Aided Des Integr Circ Syst 44(2):627–640

20. Kong Z et al (2024) PVTSizing: a TuRBO-RL-based batch-sampling optimization framework for PVT-robust analog circuit synthesis. In: ACM/IEEE design automation conference

21. Passos F et al (2018) Enhanced systematic design of a voltage controlled oscillator using a two-step optimization methodology. Integr VLSI 63:351–361

22. Mendes L, Vaz J, Passos F, Lourenço N, Martins R (2021) In-depth design space exploration of 26.5-to-29.5-GHz 65-nm CMOS low-noise amplifiers for low-footprint-and-power 5G communications using one-and-two-step design optimization. IEEE Access 9:70353–70368

23. Schulman J, Wolski F, Dhariwal P, Radford A, Klimov O (2017) Proximal policy optimization algorithms. arXiv:1707.06347

24. Towers M et al (2024) Gymnasium: a standard interface for reinforcement learning environments. arXiv:2407.17032

25. Settaluri K, Haj-Ali A, Huang Q, Hakhamaneshi K, Nikolic B (2020) AutoCkt: deep reinforcement learning of analog circuit designs. In: Design, automation & test in Europe conference & exhibition

26. Vişan C, Curăvale D, Nicolae G, Boldeanu M, Cucu H, Buzo A (2024) A novel simulations scheduler for automated circuit sizing algorithms. In: International conference on synthesis, modeling, analysis and simulation methods and applications to circuit design

27. Paszke A et al (2019) PyTorch: an imperative style, high-performance deep learning library. arXiv: 1912.01703

28. Huang S, Dossa R, Ye C, Braga J, Chakraborty D, Mehta K, Araújo J (2022) CleanRL: high-quality single-file implementations of deep reinforcement learning algorithms. J Mach Learn Res 23(274):1–18

29. Póvoa R, Lourenço N, Martins R, Canelas A, Horta N (2017) Single-stage amplifier biased by voltage combiners with gain and energy-efficiency enhancement. IEEE Trans Circ Syst II Express Briefs 65(3):266–270

30. Lourenço N, Afacan E, Martins R, Passos F, Canelas A, Póvoa R, Horta N, Dundar G (2019) Using polynomial regression and artificial neural networks for reusable analog IC sizing. In: 16th international conference on synthesis, modeling, analysis and simulation methods and applications to circuit design

Chapter 4
PVT Corner Analog IC Sizing Optimizations Boosted by ANN-Based Performance Regressors and Transfer Learning

4.1 Contributions

Several simulation-based synthesis methodologies for analog and mmWave design have successfully bypassed the simulator using machine learning (ML) models that estimate the circuit's performances [8–14]. The first reports employed ANN-based [8, 9] and neural-fuzzy-based performance estimation models [10], nonetheless, while utilizing the developed models as surrogates to the circuit simulator inside simulation-based sizing tools results in drastic speed-ups of the total end-to-end optimization time, the simplicity of the models used led to unsatisfactory prediction errors when compared with accurate off-the-shelf simulation. In [11, 12], instead of training the ANN-based performance estimation models with simulation data gathered prior to the optimization process, the simulation-based sizing loop remains unmodified for some generations and the data collected is used to train the models online (i.e., inside the optimization cycle) at a given generation. Only in subsequent generations of the optimization the developed ANN-based estimators fully replace the circuit simulator. In [13, 14], an identical concept of iteratively refining the models with online samples is applied within reinforcement learning-based automatic sizing. While all the previous approaches considered performance estimation in TT conditions only, in [6] is reported the use of specifically tailored deep ANN-based PVT regressors, however, the production of a dataset containing PVT data took more than 25 days for a single VCO with 16 corner conditions.

Balancing simultaneously all performance tradeoffs during mmWave IC design is a task that can only be achieved by EDA tools, however, PVT-inclusive synthesis incurs in prohibitive optimization times in modern EDA frameworks, pushing modern workstations' capabilities to their limits. Therefore, in this Chapter, innovative research using DL is conducted for automatic mmWave IC sizing, and its contributions are outlined as follows:

- Previous works [8–14] apply performance regressors on TT conditions only, and still, substantial errors are observed in the estimated circuit performances, hindering its use for more complex and highly non-linear mmWave circuits. In this work, deep ANN-based models tailored for accurate mmWave IC performance regression are studied and applied;
- To the best of our knowledge, for the first time in the literature, TL is established from TT to PVT schematic-level ANN-based performance estimation models, thereby avoiding the need of generating computationally expensive and time-consuming PVT datasets as in [6].
- Preceding efforts in the field apply TL-based strategies across different circuit topologies and technology nodes [15], as well as post-layout performance modelling [16, 17]. In [17], TL is applied from schematic-level ANN-based performance regressors to post-layout performance estimation models. Results obtained across several circuits show that this TL strategy can reduce the number of post-layout physical samples in more than $9\times$ compared with traditional methods.
- The proposed sizing optimization framework is tested on the automatic synthesis of three LNA topologies, designed on a 65 nm node on the challenging 28 GHz band, with complex 26-to-33-to-38-dimensional design parameter spaces, on a vast 147-dimensional performance space.

4.2   Proposed Methodology: TL-Enhanced PVT Regressors

In this section, the proposed sizing optimization framework to speed-up the PVT-inclusive synthesis of mmWave ICs is presented and detailed. Furthermore, all case studies are introduced, along with the design specifications and optimization objectives used for the demonstration of the suggested approach. Finally, a transferability analysis between TT and PVT domains is performed for the first case study, in order to ascertain the applicability of the TL strategy for the current sizing tasks.

4.2.1   TL-Enhanced PVTR Framework

The methodology proposed in this Chapter consists in the application of TL from TT to PVT corner ANN-based performance estimators, in order to speed-up the PVT-inclusive synthesis of analog and mmWave ICs. The optimization flow, as illustrated in Fig. 4.1, starts from a simulation-based sizing optimization in TT-only conditions, and a dataset is constructed, consisting in the sizing values of the candidate solutions (i.e., Table 4.1) and their respective TT performances (i.e., TT from Table 4.2). From this dataset, an ANN-based TT performance regressor is constructed, as detailed later in this Chapter, trained to estimate the circuit's TT performances from the sizing design variables. By leveraging the knowledge obtained by the TT model (i.e., the parameters of an ANN), X parallel PVT performance regressors (henceforward designated by PVTR) are created, one for each corner condition, through the reuse of the learned TT model's parameters. To ensure industry-grade accuracy in TT conditions, the TT corner is always simulated during the PVT-inclusive optimization process.

A corner optimization is then started for any specific set of performance targets, and for the first δ generations, all candidate sizing solutions are simulated across all corner conditions, to acquire accurate simulator data that will be used to fine-tune each transferred PVTR. Starting from the $\delta + 1$ generation, each PVTR becomes online within the sizing loop, estimating the respective corner performances, ultimately bypassing the need for full simulation by alleviating the overall simulator's workload, thereby significantly reducing the duration of the extensive PVT corner optimization runtimes. Given that the all PVTR are updated and used independently, a control phase analogous to the one employed in [6] may be incorporated into the sizing loop, in order to mitigate the effect of poor performance predictions, which directly affect the feasibility of the final sizing designs.

Through this TL procedure, a time-consuming PVT data acquisition phase is rendered unnecessary, in contrast to previous works employing PVT performance regressors [6], while generalizing to an arbitrary number of PVT corners, thereby eliminating the need to train separate performance estimation models for each corner, with each one being constructed during the PVT-inclusive optimization. Moreover, the solutions comprising the final pareto optimal front (POF) of the TT optimization can be used as a heuristic initial population for the corner optimization [18], preserving the natural optimization flow of analog ICs and increasing the likelihood

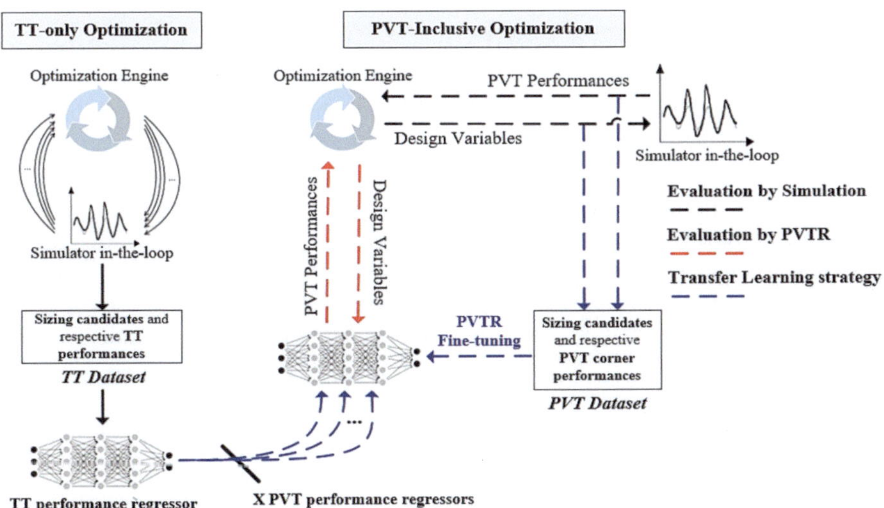

Fig. 4.1 PVT-inclusive analog/RF IC sizing optimization framework enhanced with PVT regressors and Transfer Learning

Table 4.1 Ranges/grid of the optimization variables

Device(s)	Design parameter	Units	Min	Grid	Max
L_{IN}, L_S, L_{INT}^{ε}, L_{OUT} L_{S2}^{β}, L_{OUT2}^{β}	Inner radius	μm	15	1	90
	Number of turns	–	0.5^{α}	1	5.5^{α}
	Spacing between conductors	μm	2	0.1	4
	Conductor width	μm	3	0.1	30
	Guard ring distance	μm	10	5	50
C_{IN}, C_{INT}^{*}, C_{OUT}	Length	μm	4	0.1	40
	Width	μm	4	0.1	40
M_1, M_2	Channel length	nm	60	10	240
	Channel width	μm	0.6	0.2	6
	Number of fingers	–	1	1	32
V_{G1}		V	0.4	0.01	1.2

[*] device present on both LNA2 and LNA3; [β] device only present on LNA3; [ε] device only present on LNA2; [α]for LNA2 the range of L_{IN} and L_{OUT} is 1 and 6

of steering the optimization into feasible regions of the performance and design spaces.

4.2.2 Case Studies: 28 GHz Low-Noise Amplifiers

Low-Noise Amplifiers play a critical role in modern multi-standard mmWave IC transceivers, and the proposed PVT estimators will be demonstrated on three different topologies, designed in a 65 nm CMOS node and whose schematics are illustrated in Fig. 4.2. The LNAs are based on a traditional cascode (DC coupled), and the gate voltage (VG1) of the common source transistor (M1) controls the bias. For their automatic sizing, the LNAs' components, i.e., spiral inductors L_{IN}, L_S, L_{INT} and L_{OUT}; Radio-Frequnecy (RF) MIM capacitors C_{IN}, C_{INT} and C_{OUT}; NMOS RF transistors M1 and M2; and also, bias voltage, were parameterized into 26, 33 and 38 design variables for LNA1, LNA2 and LNA3, respectively, as shown in Table 4.1. All inductive elements are components whose models are available in foundry process design kit (PDK).

4.2.3 Design Specifications and PVT Corners

For its optimization, each candidate sizing solution is simulated in TT conditions and 6 PVT corners during the sizing optimization. In each simulation performed, 21 performance values are automatically extracted, i.e., S11, S21, S22, noise figure

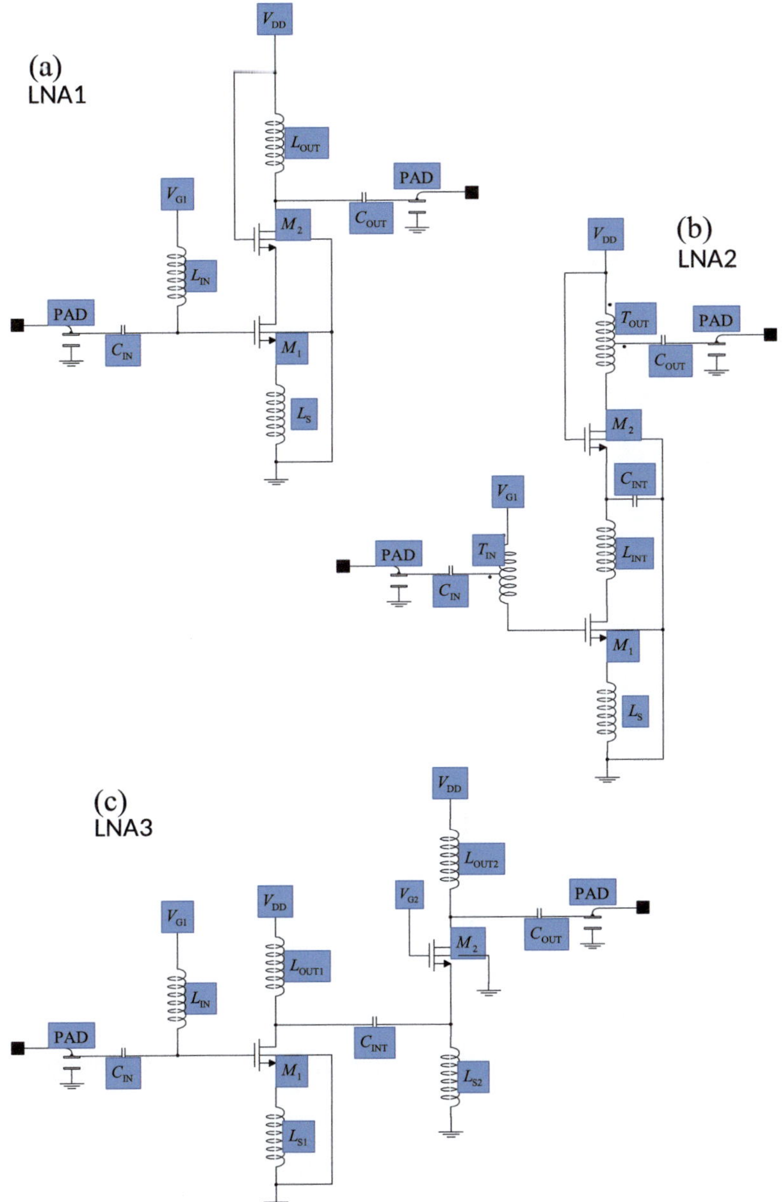

Fig. 4.2 Schematics of **a** LNA1, **b** LNA2 and **c** LNA3. VDD = 1.2 V for the LNA1 and LNA2, VDD = 0.6 V for the LNA3. Input/output signal pads were included since their shunt capacitances impact the performance

Table 4.2 Design specifications and optimization objectives

Specifications	Testbench	Units	Target
S11@26.5 GHz, 28.0 GHz, 29.5 GHz	TT, FF, SS, $V_{DD}-50$ mV, $V_{DD}+50$ mV, T0 °C, T70 °C	dB	≤-12
S21@26.5 GHz, 28.0 GHz, 29.5 GHz		dB	≥10
S22@26.5 GHz, 28.0 GHz, 29.5 GHz		dB	≤-12
NF@26.5 GHz, 28.0 GHz, 29.5 GHz		dB	≤7
flat gain up = |S21@29.5 GHz−S21@28.0 GHz|		dB	≤1.5
flat gain down = |S21@28.0 GHz−S21@26.5 GHz|		dB	≤1.5
K_f@26.5 GHz, 28.0 GHz, 29.5 GHz		–	≥1
B_{1f}@26.5 GHz, 28.0 GHz, 29.5 GHz		–	≥0
power@DC		W	≤0.05
Objectives	Testbench	Target	
S21@28.0-GHz	TT	Maximize	
NF@28.0-GHz & power@DC		Minimize	

(NF), Rollet stability factor (K_f) and the intermediate term (B_{1f}) at 26.5, 28 and 29.5 GHz frequencies, flat gain above and flat gain below 28 GHz, and lastly, DC power, as summarized in Table 4.2.

All specifications are verified in TT conditions and 6 PVT corners, consistent with the methodology followed in [18]: slow/slow {SS} and fast/fast {FF} process corners, supply voltage set to 50 mV below {$V_{DD}-50$ mV} and above {$V_{DD} + 50$ mV} standard supply, and temperature set for 0 °C {T0 °C} and 70 °C {T70 °C}. This setup ultimately results in 147 relevant performance values for the sizing problem, whose targets are summarized in Table 4.2. While all design specifications must be met during optimization, three different objectives were set: maximize S21 at 28 GHz and minimize power consumption and the NF at 28 GHz.

4.2.4 *Transferability Analysis Between TT and PVT Domains*

The transfer of knowledge between TT and PVT corner conditions can be formally characterized as homogenous transfer learning (HTL) [6, 19]. In this setting, the source (*S*) and target domains (*T*) share the same identical feature space ($X_S = X_T$) and similar label spaces ($Y_S \approx Y_T$), differing only in their underlying data distributions. Figure 4.3 depicts the distributions of the three optimized performance figures (described in Table 4.2) across the typical and PVT corners described in the previous sections for the LNA1. The data was obtained through a combined TT and PVT corner optimization, starting from a randomly sampled population and executed using the sizing tool presented in [20], using a population size of 100 elements optimized through 1000 generations.

By analyzing the resulting distributions, it is clear that all performance figures exhibit similar marginal behaviors across all PVT corners (closely related mean and median values), indicating that the conditional relationships $P_S(Y|X)$ and $P_T(Y|X)$

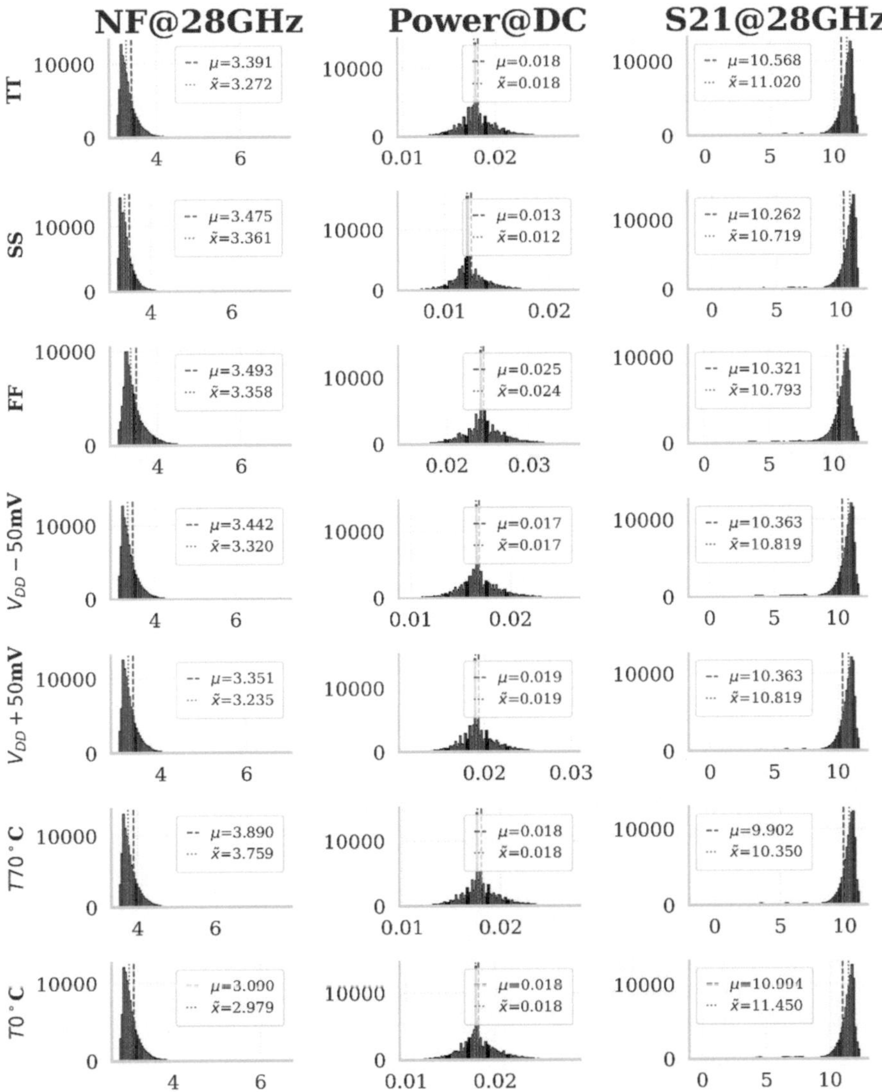

Fig. 4.3 Performance distributions of S21@28 GHz, NF@28 GHz and power@DC in TT and PVT corners for the LNA1

are closely aligned. This supports the use of HTL strategies, like fine-tuning the pre-trained TT model with a small number of PVT samples, to achieve effective adaptation. Consequently, the predictive model trained at the TT corner is expected to generalize effectively to other PVT corners with minimal retraining effort.

4.3 Development of the TT Performance Regressors

This section begins by introducing the general architecture considered for the TT performance regressors. It then provides a detailed description of the datasets used in their development. Furthermore, all the steps undertaken for constructing the LNA3 TT regressor are described, which were also employed for the development of the other TT regressors. Finally, this section concludes with a presentation of all the model configurations, which will be used to evaluate the methodology detailed in Sect. 4.2.1.

4.3.1 General In/Out Structure

The general In/Out structure of all TT performance regressors is depicted in Fig. 4.4. In all regression models, an architecture consisting of 3 fully connected hidden layers without weight sharing was selected to carry out the TL strategy. The input layer of each network will be comprised of the sizing design variables, i.e., the optimization variables of Table 4.1, and the output layers will have 21 units, i.e., the TT performance figures presented in Table 4.2.

4.3.2 Datasets

In order to generate the datasets needed for constructing all TT performance regressors, 3 TT-only optimizations were carried out for each LNA, using a population size of 100 elements optimized through 1000 generations, using the simulation-based sizing tool of [20], culminating in 100′100 data entries for each circuit, with each sample representing a given sizing candidate solution and its TT performances obtained through simulation. These optimizations provided 100 non-dominated solutions in the final POF for both the LNA1 and LNA2, and 57 non-dominated solutions for the LNA3. Due to the differing ranges of the sizing design variables and performance figures (Tables 4.1 and 4.2, respectively), data preprocessing is required to improve the quality of the datasets used for ANN training, since discrepancies in the orders of magnitude can lead to incorrect weight updates and issues such as vanishing or exploding gradients during the backpropagation update [21]. To mitigate these effects, two preprocessing methods are commonly considered: normalization (4.1)

Table 4.3 Dimensions of the datasets used to train the TT regressor

Circuit	Training samples	%Total
LNA1	84′ 466	84.38
LNA2	83′ 805	83.72
LNA3	84′ 103	84.02

and standardization (4.2).

$$x_{\text{normalized [a,b]}} = \frac{x - x_{\min}}{x_{\max} - x_{\min}} \cdot (b - a) + a \tag{4.1}$$

$$x_{\text{standardized}} = \frac{x - \mu}{\sigma} \tag{4.2}$$

Normalization scales the values into a desired fixed range, and although this transformation does not change the features' distributions, it enhances the effects of outliers [22], which correspond to observations of a certain variable that lie abnormally distant from the other data points. On the other hand, standardization scales the values considering the mean (μ) and standard deviation (σ). Considering that the mean absolute percentage error (MAPE) (4.5) will be used as quality measure of the models, data normalization into the range [1, 2] was performed in all datasets, in order to prevent the MAPE values from becoming excessively large when considering near-zero values. To deal with the influence of outliers, a statistical approach was used, with all data rows containing the top or bottom 1% values of all performance figures being completely eliminated from the respective datasets. Likewise, all duplicate rows in each dataset were also discarded. Table 4.3 shows the number of training samples used to develop each TT regressor, after application of the mentioned data processing techniques.

With the datasets already processed and ready to be used, each one was randomly split into 70% train, 20% validation and 10% test. In the following section, the experiments conducted to build LNA3's TT regressor are presented and discussed. At the end of this section, all final model configurations are detailed, which followed the same tuning strategy.

4.3.3   Hyperparameter Tuning

To assess each combination of hyperparameters during the tuning phase, two metrics commonly used for evaluating regression models were employed: the mean absolute error (MAE) (4.4) and the MAPE (4.5). The former measures the average absolute deviation between the true value y_i and the predicted output $\hat{y}_i$, while the latter quantifies the average percentage deviation. In all tested configurations, the mean squared error (MSE) (4.3) was adopted as the loss function, as it is a standard choice for a variety of regression tasks [21]. Furthermore, ADAM [23] was chosen as the optimizer in all developed networks.

$$MSE = \frac{1}{n} \sum_{i=1}^{n} \left(y_i - \widehat{y}_i\right)^2 \tag{4.3}$$

$$MAE = \frac{1}{n} \sum_{i=1}^{n} \left|y_i - \widehat{y}_i\right| \tag{4.4}$$

$$MAPE = \frac{1}{n} \sum_{i=1}^{n} \left|\frac{y_i - \widehat{y}_i}{y_i}\right| \times 100\% \tag{4.5}$$

The first step in the hyperparameter tuning phase is to tune the number of units per hidden layer, considering three fully connected hidden layers, and the learning rate. All the models presented in Table 4.4 were trained using the Keras [24] and Tensorflow [25] libraries of the Python programming language, with fixed parameters such as the batch size (128) and activation function of the hidden layers (ReLU). Although several configurations were tested, the results presented represent the best configurations that were explored, with the values highlighted corresponding to the best values found in all evaluation metrics. To determine the best performing configuration, the following values were considered:

- Size of hidden layer 1: [400, 440, 480]
- Size of hidden layer 2: [280, 320, 380]
- Size of hidden layer 3: [120, 160, 200]
- Learning rate: [0.0001, 0.0005, 0.001]

Table 4.4 Results of hidden layers size and learning rate tuning study. Activation function: ReLU, batch size: 128, dropout rate: 20%

Hidden layer sizes			Learning rate	Train			Validation		
#1	#2	#3		MSE ($\times 10^{-3}$)	MAE ($\times 10^{-2}$)	MAPE	MSE ($\times 10^{-3}$)	MAE ($\times 10^{-2}$)	MAPE
480	380	120	0.0001	3.021	3.116	2.162	2.502	2.311	1.602
480	380	120	0.0005	2.563	2.622	1.798	1.312	1.822	1.258
480	380	120	0.001	2.189	2.881	1.953	2.467	2.456	1.651
480	380	160	0.0001	3.442	3.466	2.079	2.219	2.278	1.519
480	380	160	0.0005	**2.014**	2.521	1.767	**1.211**	**1.803**	**1.257**
480	380	160	0.001	2.081	2.714	1.878	1.991	2.198	1.452
480	380	200	0.0001	2.133	3.267	2.037	2.123	2.256	1.528
480	380	200	0.0005	2.114	**2.512**	**1.725**	1.881	2.072	1.401
480	380	200	0.001	2.227	2.790	1.856	2.124	2.568	1.711

Based on the results obtained across all trained networks, an architecture consisting of 480, 380 and 160 units in each hidden layer was considered for the remaining phases of the hyperparameter tuning process. Table 4.5 shows the values obtained for the activation function study. Since Leaky ReLU outperformed all other tested activation functions (besides the MSE in the training set), it was chosen as the final activation function of the hidden layers, and was used to assess the best performing values of the remaining hyperparameters.

Table 4.6 shows the results of the batch size tuning study. Due to the values obtained, a batch size of 128 was selected for the final model. Table 4.7 depicts the results derived from the study of the dropout rate applied to all hidden layers. The results clearly indicate that the use of dropout decreases the overall performance of the trained networks, therefore no dropout layers were added to the final LNA3's TT regressor.

Table 4.5 Results of activation function study, dropout rate: 20%, batch size: 128

Activation function	Train			Validation		
	MSE $(\times 10^{-3})$	MAE $(\times 10^{-2})$	MAPE	MSE $(\times 10^{-3})$	MAE $(\times 10^{-2})$	MAPE
ReLU	**2.014**	2.521	1.767	1.211	1.803	1.257
Leaky ReLU	2.083	**2.446**	**1.687**	**0.988**	**1.609**	**1.132**
ELU	2.124	2.561	1.729	1.001	1.840	1.226
Sigmoid	4.022	3.920	2.708	3.326	3.127	2.089

Table 4.6 Results of batch size study, dropout rate: 20%

Batch size	Train			Validation		
	MSE $(\times 10^{-3})$	MAE $(\times 10^{-2})$	MAPE	MSE $(\times 10^{-3})$	MAE $(\times 10^{-2})$	MAPE
32	2.171	2.512	1.693	1.011	1.612	1.141
64	**2.081**	2.455	1.687	1.003	**1.608**	1.136
128	2.083	**2.446**	**1.687**	**0.988**	1.609	**1.132**
256	2.341	2.599	1.695	1.104	1.811	1.234

Table 4.7 Results of dropout rate study

Dropout rate	Train			Validation		
	MSE $(\times 10^{-3})$	MAE $(\times 10^{-2})$	MAPE	MSE $(\times 10^{-3})$	MAE $(\times 10^{-2})$	MAPE
0%	**0.412**	**1.212**	**0.838**	**0.544**	**1.310**	**0.914**
5%	0.944	1.763	1.217	0.836	1.472	1.021
10%	1.201	2.010	1.390	0.765	1.449	1.009
20%	2.083	2.446	1.687	0.988	1.609	1.132

4.3.4   Evaluation and Final Model Configurations

Table 4.8 presents the final model architectures and training settings obtained for all final TT regressors, outlining the diverse set of hyperparameters selected for all final models.

Figures 4.5, 4.6 and 4.7 present the final model losses obtained for the LNA1, LNA2 and LNA3's TT regressors, respectively

To evaluate each assembled TT performance estimator, the test sets generated in Sect. 4.3.2 were used. Table 4.9 shows the values obtained in each test set. These

Table 4.8 Final architecture and training settings of each TT regressor

Hyperparameter	LNA1	LNA2	LNA3
Hidden layer units	480, 350, 220	480, 350, 220	480, 380, 160
Activation function	ReLU	ReLU	Leaky ReLU
Learning rate	0.0003	0.0003	0.0005
Batch size	128	128	128
Dropout rate	20%	20%	0%
Number of epochs	300	300	300

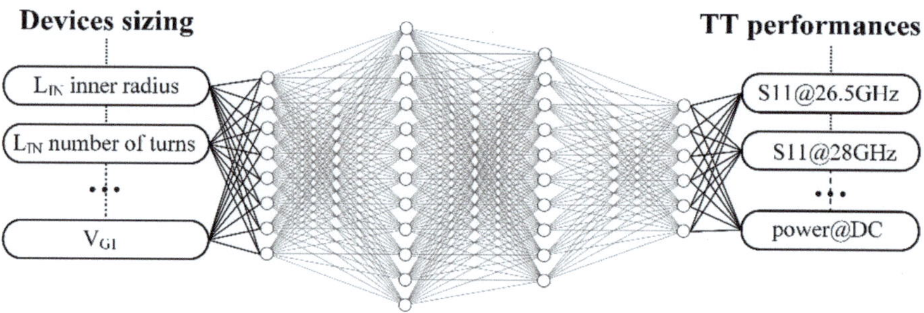

Fig. 4.4 General In/Out structure of each TT performance regressor

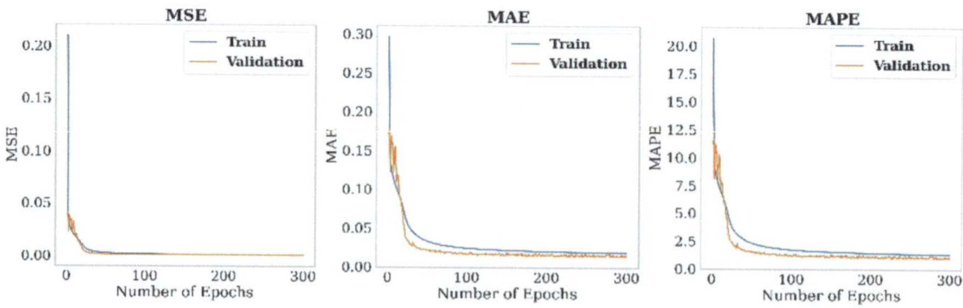

Fig. 4.5 Final model losses for LNA1's TT regressor

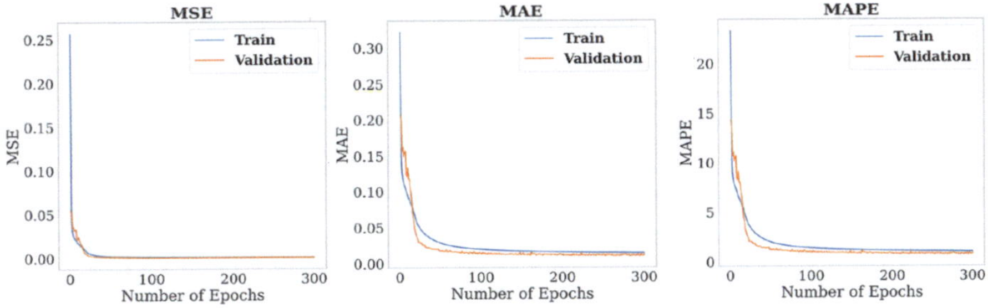

Fig. 4.6 Final model losses for LNA2's TT regressor

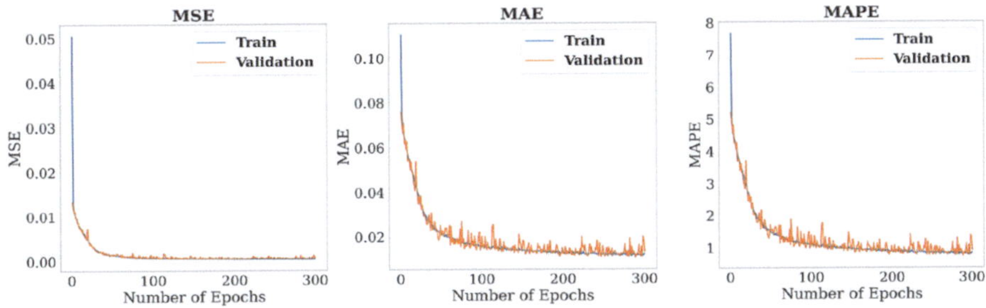

Fig. 4.7 Final model losses for LNA3's TT regressor

Table 4.9 Results on the test set for all TT regressors

Circuit	MSE ($\times 10^{-3}$)	MAE ($\times 10^{-2}$)	MAPE
LNA1	0.7759	1.1759	0.9362
LNA2	0.4972	1.0247	0.8763
LNA3	0.5436	1.3198	0.9187

results show that the tuned networks are capable of accurate estimations, setting an interesting baseline for the TL procedure.

4.4 Optimization Results

This section presents the experimental results obtained using the proposed sizing optimization framework detailed in Sect. 4.2.1, with the goal of validating its effectiveness through multiple optimization runs across all circuits described in Sect. 4.2.2. The experiments were designed to assess the framework's generalization capability and scalability with increasing circuit complexity, characterized by design spaces of 26, 33 and 38 optimization parameters for LNA1, LNA2 and LNA3, respectively, in a vast 147-dimensional performance space. As an initial evaluation, a brute-force approach was explored, in which the PVT estimators fully replaced the circuit simulator within the sizing loop for both LNA1 and LNA2, with each PVTR estimating the corresponding corner performances in all generations following generation δ. Although this setup resulted in substantial speed-ups, the quality of the obtained solutions may limit its applicability to more complex circuit topologies. To address this limitation, a hybrid strategy was subsequently explored, where each fine-tuned PVTR operated collaboratively with the circuit simulator inside the optimization cycle, rather than completely replacing it. The proposed methodology introduced in the previous section was incorporated in the simulation-based sizing tool of [20], and developed using the Keras and Tensorflow libraries of the Python programming language.

4.4.1 TL-Enhanced PVTR Framework as a Full Simulator Surrogate

With the TT performance regressors for the LNA1, LNA2 and LNA3 already tuned to achieve good accuracy, and considering the PVT-inclusive optimization framework previously described, the initial trials conducted for both LNA1 and LNA2 circuit topologies involved fully replacing the simulator within the optimization cycle after a given generation. This replacement occurred only after a mandatory number of full-simulation generations had been completed to provide a sufficient amount of simulator-grade PVT data to fine-tune the transferred TT performance regressors. This generation is hereafter referred to as the fine-tuning generation (δ). To evaluate the effectiveness of the proposed approach and the TL procedure, two optimization runs were performed for the LNA1 and LNA2 circuit topologies, using the design specifications presented in Table 4.2. Each run employed a population size of 100 individuals and was executed for 500 generations.

One of the main challenges of this approach lies in the fact that the fine-tuning process is performed online, i.e., without developer intervention. This autonomous adaptation may lead to issues such as underfitting, overfitting, or other behaviors that hinder the desired generalization of the models. Choosing an appropriate fine-tuning strategy is therefore essential to achieve the desired accuracy of each PVTR, which directly impacts the reliability of the corner performance estimations. To address this challenge, bayesian optimization (BO) was employed as the automatic fine-tuning method for these first experiments, with the retraining process occurring for a maximum of 100 epochs, and early stopping with a patience set to 10 epochs is applied to prevent overfitting and save computational resources. For this BO-assisted TT to PVT adaptation, a maximum number of 10 evaluations was adopted.

Since knowing how many parameters of the pre-trained model need to be tuned to achieve the best overall performance is not a trivial task, the main hyperparameters considered in the tuning process were: the number of layers to be updated (i.e., which layers of the TT model need to be updated for a successful adaptation to each

Table 4.10 Number of fine-tuning samples for each PVTR

Circuit	SS	FF	$V_{DD}-50$mV	$V_{DD}+50$mV	T0°C	T70°C
LNA1	8322	8401	8352	8358	83,525	8356
LNA2	8290	8277	8263	8261	8270	8256

corner's performance space) and the learning rate, which affects the stability and generalization of the tuned ANN. Both parameters are passed as arguments to the BO-assisted fine-tuning procedure. In these optimization runs, BO finds the optimal number of layers to be updated, and explores a linear subspace of learning rates centered at the value of the original learning rate (i.e., the one used to train the TT model). Table 4.10 depicts the number of fine-tuning samples used to train each PVTR, after application of the same data processing techniques that were used in the initial TT dataset.

To enable a comprehensive comparison between the quality of the solutions obtained using the TL-enhanced PVTR framework and those derived from the original sizing tool of [20], two additional corner optimization runs were performed using the unmodified optimization loop, under the same optimization settings for both LNA1 and LNA2. Figures 4.8 and 4.9 present the final POF obtained, illustrating the tradeoffs between S21@28 GHz and power consumption, obtained with the modified optimization cycle (incorporating the PVTR framework) and with the baseline sizing tool (without PVTR).

A detailed analysis of the final design tradeoffs obtained with both the original and the modified sizing tools reveals a clear performance contrast between the two LNA optimizations. The LNA2 optimization using the PVTR enhanced framework produced 86 non-dominated solutions in the final POF, whereas the LNA1 optimization yielded only a single non-dominated solution. This isolated solution exhibited a substantially lower S21@28 GHz value compared to all other solutions found by the original framework, although it achieved a power consumption superior to 7 out

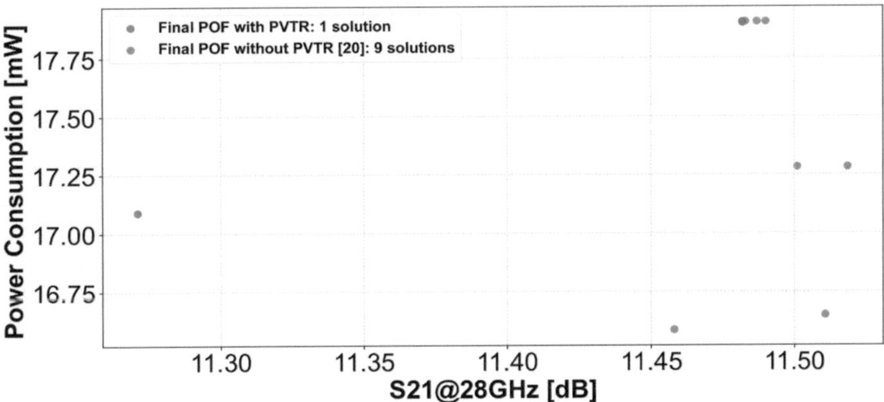

Fig. 4.8 Final POF (tradeoff S21@28.0 GHz and Power@DC) for the LNA1

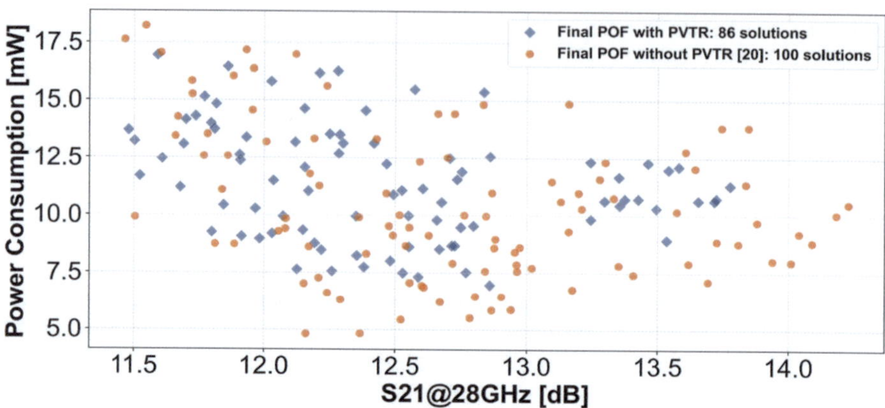

Fig. 4.9 Final POF (tradeoff S21@28.0 GHz and Power@DC) for the LNA2

of the 9 baseline designs. This outcome suggests that, for the LNA1, the fine-tuning process of the PVTR may not have gathered sufficient data within that region of the performance space, leading to inaccurate performance estimations.

Conversely, the enhanced optimization of the LNA2 demonstrated strong consistency, generating 86 non-dominated solutions with well-balanced tradeoffs across competing objectives. The extreme values of these non-dominated solutions, obtained with the proposed framework for both LNA1 and LNA2, are summarized in Table 4.11, as well as the extreme values found in the final POF of the TT optimizations (the ones conducted to generate the datasets with which the TT regressors were built). To assess the feasibility of the non-dominated solutions produced by the proposed approach, these were subsequently simulated under the complete set of PVT corner conditions. This verification step confirmed one feasible solution for LNA1 (i.e., fulfilling all design requirements listed in Table 4.2) and 86 feasible solutions for the LNA2. All non-dominated solutions found using the original sizing framework reached feasibility.

Using an AMD Ryzen 3960X workstation, the unmodified PVT-inclusive optimizations took approximately 65 and 64 h for LNA1 and LNA2, respectively. The proposed framework was able to achieve competitive sizing results in less than 19 h for both circuits, reducing the simulator's workload down by approximately 68.5%, translating into speed-up factor of around 3.4× compared to the original sizing tool, a substantial speed-up, especially when allied to the fact that it only required 13 h of full-simulation synthesis (i.e., the total time required to evaluate all sizing candidates throughout the first δ generations, with the original and unmodified sizing loop) to collect enough data to build reliable PVT performance estimators, culminating in a dataset generation effort decrease of approximately 83% relative to the approach employed in [6].

Table 4.11 Minimum and maximum POF values found for the TT and TL-Enhaced PVT-inclusive optimizations for the LNA1 and LNA2

Final POF values		LNA1		LNA2	
		Opt. TT	Opt. PVT[1]	Opt. TT	Opt. PVT[1]
#non-dominated solutions		100	1	100	86
S21@28.0GHz (dB)	Min	10.68	11.27	10.57	11.48
	Max	11.23		14.66	13.78
NF@28.0GHz (dB)	Min	3.09	3.12	2.89	3.13
	Max	3.62		4.67	4.37
power@DC (mW)	Min	13.51	17.09	2.19	7.00
	Max	20.37		16.91	16.90

[1] Opt. TT used as the initial population

4.4.2 Controlled TL-Enhanced PVTR Framework

Considering the previously discussed issues, particularly those observed in the LNA1 optimization, one way to mitigate the impact of inaccurate performance estimations during the optimization process is to introduce a control phase in the sizing loop. In this phase, for each generation following the fine-tuning generation, 30% of the sizing solutions in the current generation are simulated across all corner conditions and compared against the corresponding PVTR estimations. If the discrepancy between simulated and estimated results remains below a predefined threshold error (Δ), the corresponding PVTR can be considered reliable for that region of the design space and may thus be utilized to evaluate the remaining solutions in that generation. Although this modification may not achieve the same level of computational speed-up, due to the increased simulation costs, it is expected to significantly enhance the quality and feasibility of the final sizing designs. For the following experiments, a threshold error of $\Delta = 10\%$ was adopted, a value slightly higher than the one employed in [6] due to the incapacity of incorporating the TT performances in the input layer of each network. The generation flow (for every generation following δ) of the controlled PVT-incluisive optimization is depicted in Fig. 4.10.

4.4.2.1 Optimization Results for the LNA1

To validate the effectiveness of the controlled PVT-inclusive optimization approach, two additional optimization runs were carried out for the LNA1, using the same design specifications presented in Table 4.2 and relaxing the S21@26.5 GHz and S21@29.5 GHz targets with a soft constraint of 33%. For consistency and fair comparison, the fine-tuning generation was again set to 100. To further assess the overall speed-up granted by the proposed methodology, the total number of generations in these runs was set to 500 and 1000, respectively, while maintaining the

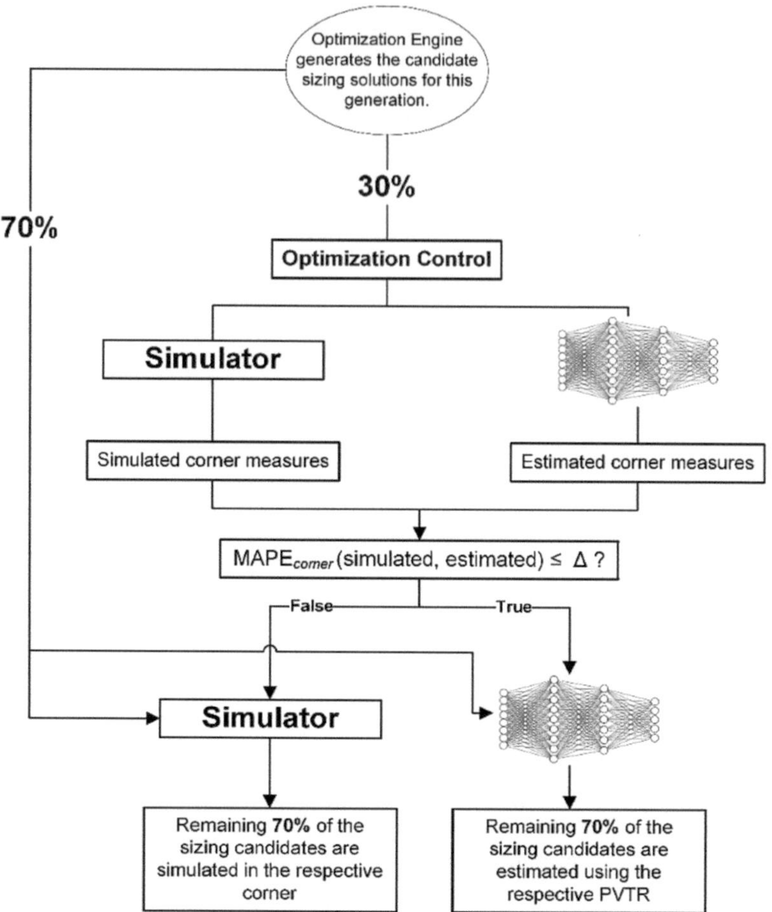

Fig. 4.10 Generation flow of the controlled TL-enhanced PVTR framework

same population size, in order to attest its capabilities in longer optimization runs, where the likelihood of finding feasible solutions is increased. Figures 4.11 and 4.12 exhibit the percentage of the total PVT corner evaluations fully bypassed by the corresponding PVTR throughout the optimization processes employing 500 and 1000 generations, respectively.

Figure 4.13 presents the final design tradeoff values obtained for the LNA1 across the conducted optimization runs. The TL-enhanced PVTR models exhibited robust generalization across different regions of the design and performance spaces, maintaining a comparable proportion of successfully predicted evaluations. It is evident that running longer optimizations yields better extreme values for the optimized performance figures, with the final results obtained with the optimization employing 1000 generations demonstrating improved values for S21@28GHz, power@DC and NF@28GHz. All solutions in the final POF obtained using 1000 generations were

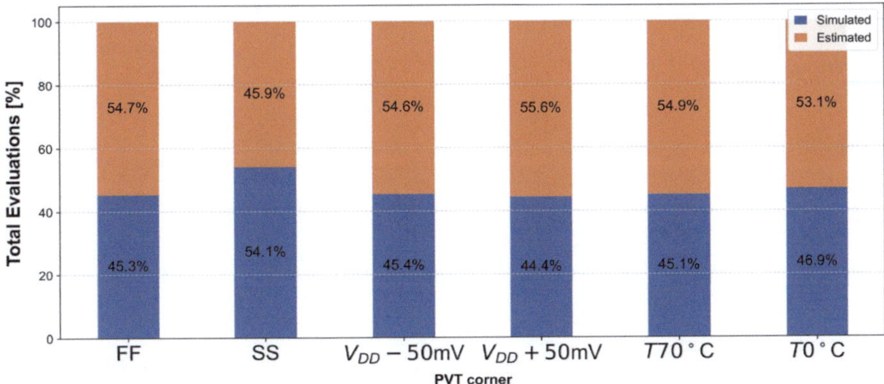

Fig. 4.11 Number of evaluations simulated and estimated for the LNA1 controlled PVT-inclusive optimization using 500 generations

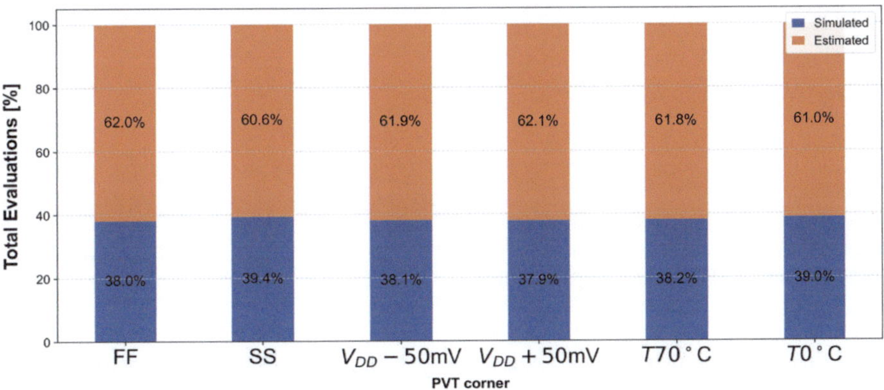

Fig. 4.12 Number of evaluations simulated and estimated for the LNA1 controlled PVT-inclusive optimization using 1000 generations

successfully validated through off-the-shelf simulation across all PVT corner conditions, confirming their feasibility. Figure 4.14 depicts the fine-tuning losses obtained through the BO-assisted fine-tuning procedure in the optimization process employing 1000 generations, with the horizontal axis representing the number of epochs in the retraining process of each transferred PVTR.

It is evident that the obtained results closely align with those of the initial pre-trained TT regressor, suggesting effective knowledge transfer. The plots presented for the {SS} corner might be indicative of overfitting in the target model, nonetheless, considering the loss values in all evaluation metrics, it should not affect the PVTR's predictive capabilities, as shown in Fig. 4.12, presenting a consistent number of successful evaluations compared to the other fine-tuned networks. The controlled PVT-inclusive optimization was able to reduce the total optimization time from

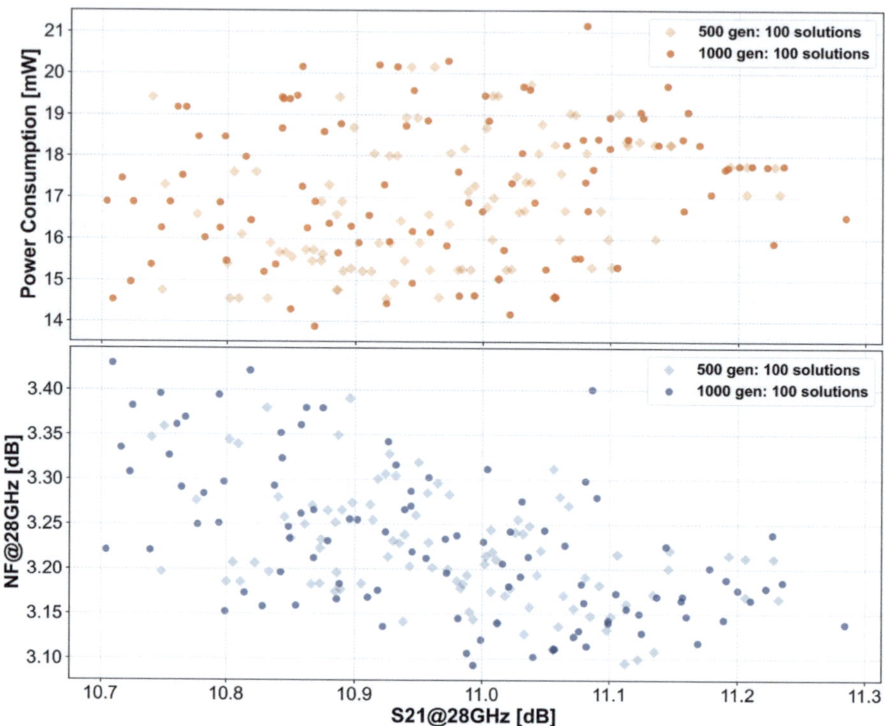

Fig. 4.13 Comparison of the final tradeoffs S21@28 GHz vs power@DC and NF@28 GHz for the LNA1 controlled PVT-inclusive optimizations

128.71 h (using the original sizing tool) to 55.14 h in the 1000-generation run, representing a speed-up factor of around 2.33×, reducing the overall simulator workload down by approximately 52.8% (given by the total number of PVT corner evaluations bypassed by the corresponding PVTR and accounting for the simulation of the TT corner through the course of the corner optimization).

4.4.2.2   Optimization Results for the LNA2

Considering the results obtained in the previous conducted experiments, the same methodology was employed to the automatic PVT-aware synthesis of the LNA2, again using the design specifications detailed in Table 4.2. However, for this circuit, the BO-assisted fine-tuning procedure failed to generate PVTR models capable of producing performance estimates with errors below the predefined threshold error Δ. This outcome likely happened due to the increased dimensional parameter space of the LNA2, which can be explained by BO's poor performance in relatively high dimensional spaces [26]. Instead of relying on a probabilistic search of the hyper-parameter space, a deterministic approach was leveraged, similar to the fine-tuning

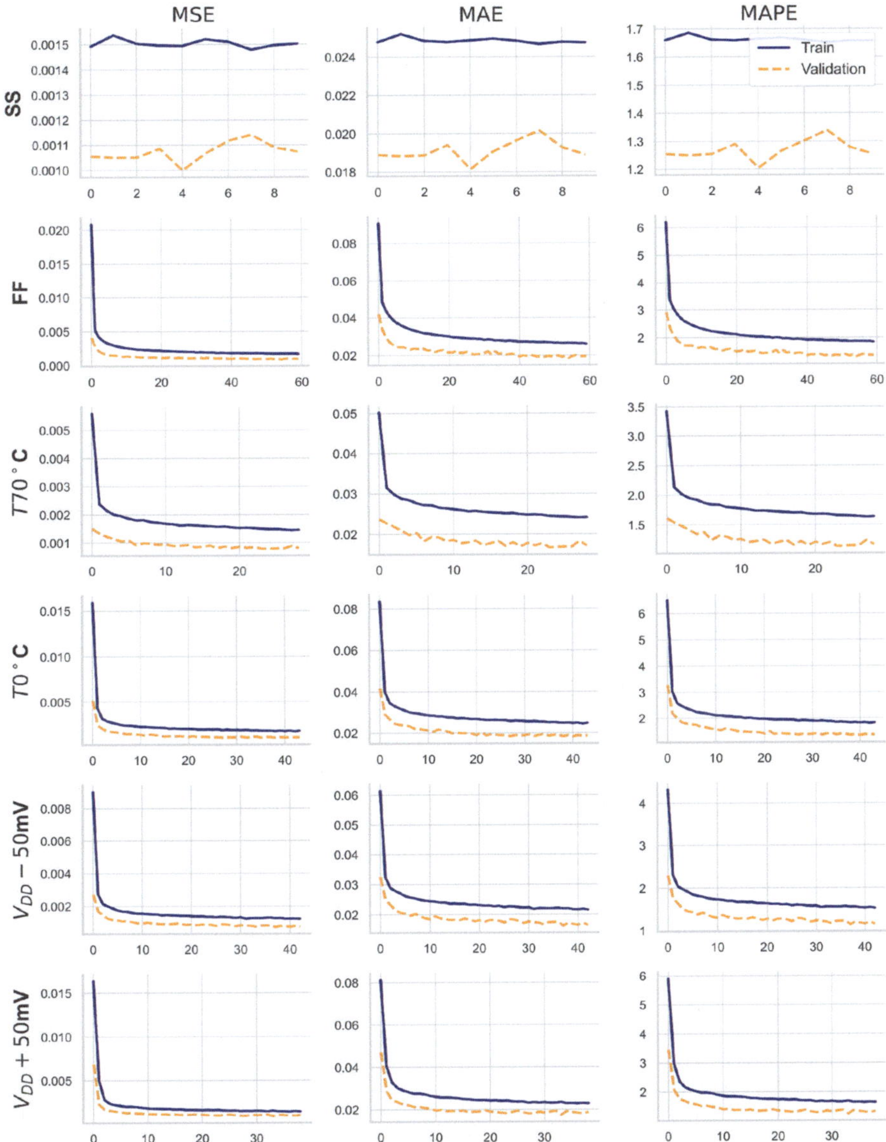

Fig. 4.14 Fine-tuning losses for the LNA1

method used in [17] for constructing post-layout performance estimation models. For their accurate adaptation, the fine-tuning process was performed using 200 epochs, and a light dropout rate of 5% was applied to all hidden layers for the retraining process.

Figure 4.15 shows the final design tradeoffs between S21@28 GHz, power@DC and NF@28 GHz obtained in the 500 and 1000-generation optimization runs. Similar to the results obtained for the LNA1, the optimization process using 1000 generations was again able to produce sizing designs with better performance targets in terms of S21@28 GHz and NF@28 GHz. Figure 4.16 shows the total percentage of PVT corner evaluations successfully estimated by each PVTR in the 1000-generation run. When compared with the results obtained in the controlled optimization run for the LNA1 under identical settings, it is clear that the fine-tuned networks produced a significantly less number of estimations. As depicted in Table 4.12, a detailed analysis of the mean (μ) and median ($\tilde{x}$) errors made by each PVTR in the control phase of the optimization process shows that these values are closely aligned with the threshold value adopted for Δ, suggesting that the predictive models operated near their tolerance limit. Although increasing the value of Δ would, in practice, increase the computational speed-up, it would also compromise the reliability of the corner performance estimations.

After accurately simulating all non-dominated solutions obtained in the final POF of these optimization runs across all PVT corner conditions, a total of 81 and 97

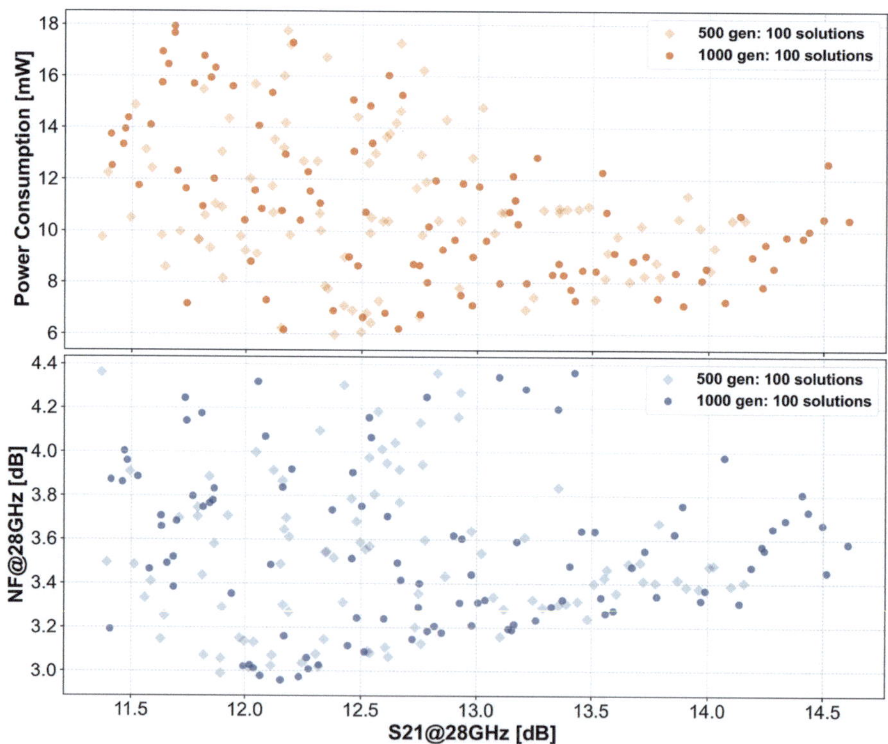

Fig. 4.15 Comparison of the final tradeoffs S21@28 GHz vs power@DC and NF@28 GHz for the LNA1 controlled PVT-inclusive optimizations

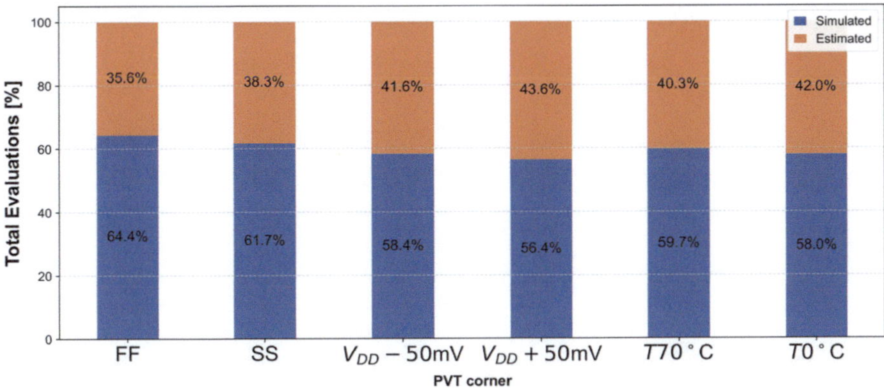

Fig. 4.16 Number of evaluations simulated and estimated for the LNA2 controlled PVT-inclusive optimization using 1000 generations

Table 4.12 Fine-tuning samples and errors made by each PVTR in the LNA2 controlled PVT-inclusive optimization using 1000 generations

	SS	FF V_{DD}-50mV V_{DD+}50mV T0ºC T70ºC
Fine-tuning samples	8322	8401 8322 8358 8355 8356
$\mu_{error}(\%)$	11.86	10.99 12.67 10.67 10.84 14.10
$\sigma_{error}(\%)$	13.01	14.22 20.18 16.02 14.52 15.41
$\tilde{x}_{error}(\%)$	9.48	9.29 8.67 8.41 9.07 8.88

feasible solutions were identified for the 500 and 1000-generation runs, respectively. Among the 19 infeasible solutions from the 500-generation run, 13 violated a single design specification, while 6 failed to satisfy two design specifications. In contrast, of the 3 infeasible solutions obtained in the 1000-generation run, each failed to meet only one of the design requirements listed in Table 4.2. Figure 4.17 showcases all fine-tuning losses obtained in 1000-generation run. These results demonstrate effective adaptation to each corner's performance spaces, with all fine-tuned models showing MAPE values in both train and validation sets below 1%. The proposed controlled sizing framework was able to alleviate the total simulator workload down by approximately 34.5%, translating into a speed-up factor of 2.06×, consequently decreasing the overall optimization duration from 127.33 h (using the baseline sizing tool) to just 61.65 h.

4.4.2.3 Optimization Results for the LNA3

Drawing on the results obtained for the LNA1 and LNA2 controlled optimizations, this section further validates the proposed sizing methodology, highlighting its ability to accelerate PVT-inclusive optimizations in even more complex circuits, such as the LNA3. Identically to the experiments conducted for the previous circuits, an

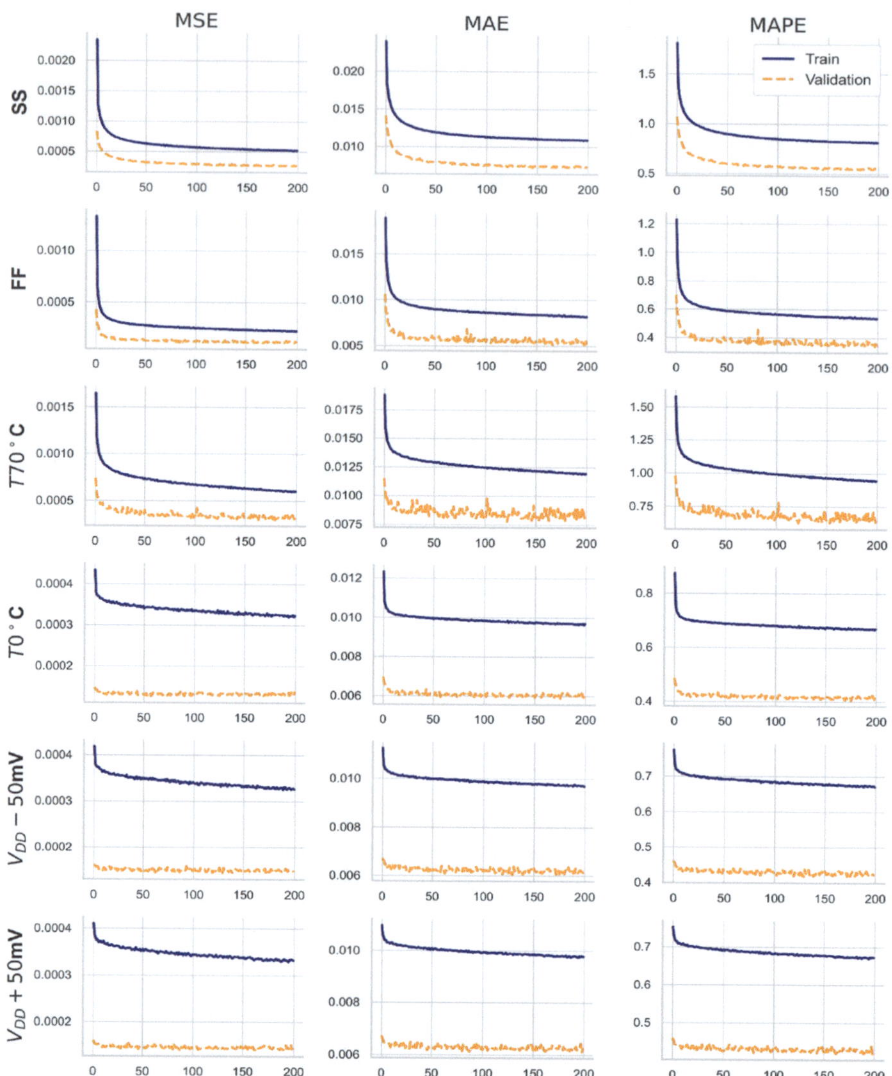

Fig. 4.17   Fine-tuning losses for the LNA2

optimization run using a population size of 100 elements optimized through 1000 generations was carried out for the LNA3. Figure 4.18 shows the percentage of total corner evaluations proficiently evaluated using the fine-tuned PVTR models. These results demonstrate similar outcomes than the one's obtained for the controlled PVT-inclusive optimization of the LNA1 under the same optimization settings. Table 4.13 showcases the predictive errors made by each independent PVTR in the control phase of the optimization, as well as the number of processed samples used to fine-tune the previously developed LNA3's TT regressor across all PVT corners.

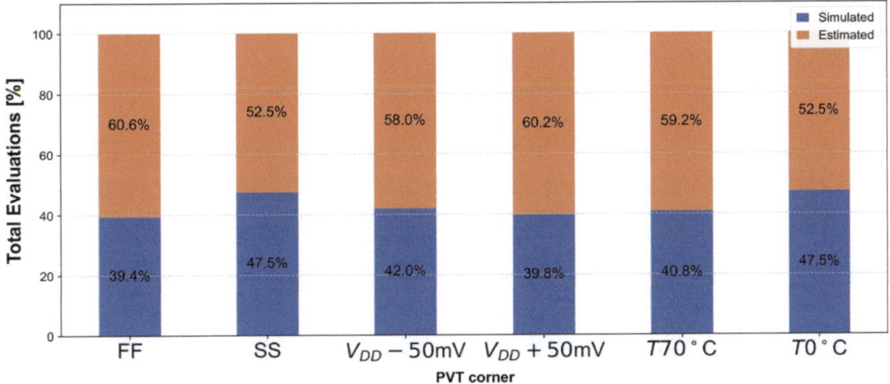

Fig. 4.18 Number of evaluations simulated and estimated for the LNA3 controlled PVT-inclusive optimization

Table 4.13 Fine-tuning samples and errors made by each PVTR in the LNA3 controlled PVT-inclusive optimization

	SS	FF V_{DD}-50mV V_{DD}+50mV T0°C T70°C
Fine-tuning samples	8249	8281 8262 8255 8243 8278
$\mu_{error}(\%)$	5.17	8.36 6.01 5.67 6.67 8.72
$\sigma_{error}(\%)$	6.51	7.25 10.10 3.14 4.49 11.26
$\tilde{x}_{error}(\%)$	4.06	6.88 4.49 4.90 4.68 6.59

Figure 4.19 illustrates the evolution of the POF throughout the LNA3 controlled PVT-inclusive sizing process, showing the non-dominated solutions obtained at generations 100, 500 and 1000. In contrast to the previous controlled PVT-inclusive optimizations for the LNA1 and LNA2, this optimization failed to produce more than one non-dominated solution in the final POF. The targets of this isolated solution are presented in Table 4.14, along with the extreme values found in the original TT POF.

This bottleneck might be explained by the fact that the initial TT optimization failed to produce the intended 100 solutions in its final POF. Increasing the total number of generations in the TT optimization would likely mitigate this limitation by improving the quality and diversity of the final TT POF. After accurately simulating the final non-dominated solution in all corner testbenches resulted in a sizing design that fails to meet 3 performance requirements, out of all those imposed in Table 4.2. Figure 4.20 illustrates the fine-tuning losses obtained in this optimization process. As can be examined, all PVTR were able to achieve MAPE values below th 1% mark in both train and validation sets, suggesting that the selected fine-tuning procedure can in fact produce accurate models inside the optimization cycle.

Considering the evaluations proficiently made by the performance estimators during the PVT-aware optimization, the developed framework reduced the overall simulator's workload down by approximately 55.80%, decreasing the end-to-end

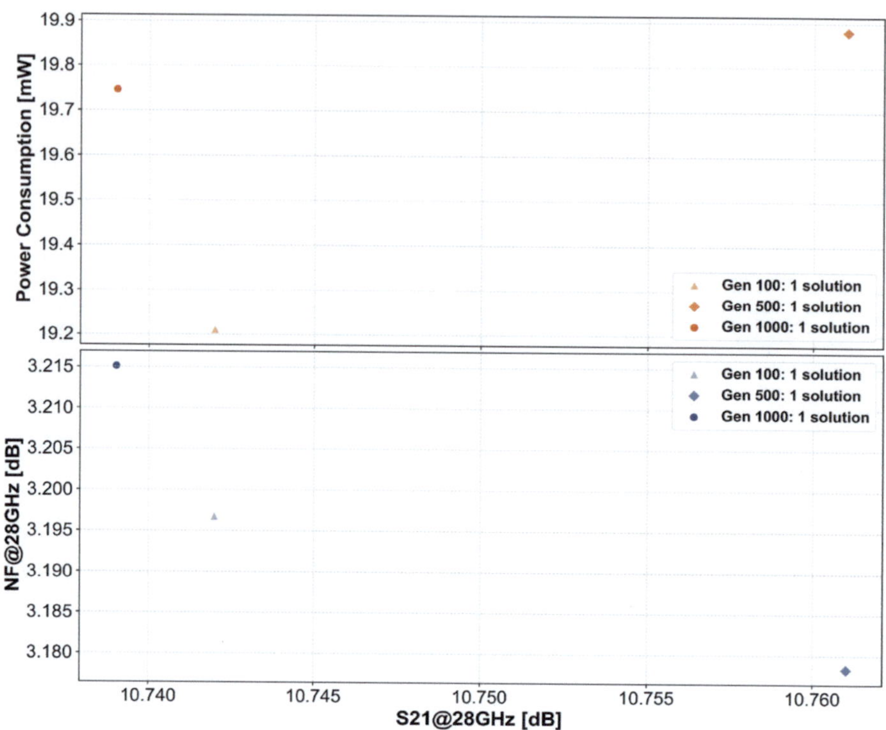

Fig. 4.19 Final tradeoffs S21@28GHz vs power@DC and NF@28GHz for the LNA3 controlled PVT-inclusive optimization

Table 4.14 Minimum and maximum POF values found for the TT and controlled PVT-inclusive optimizations for the LNA3

Final POF Values		LNA3	
		Opt. TT	Opt. PVT[1]
#non-dominated solutions		57	1
S21@28.0GHz (dB)	Min	10.69	10.74
	Max	11.02	
NF@28.0GHz (dB)	Min	3.15	3.22
	Max	3.41	
power@DC (mW)	Min	18.59	19.75
	Max	22.11	

[1]Opt. TT used as the initial population

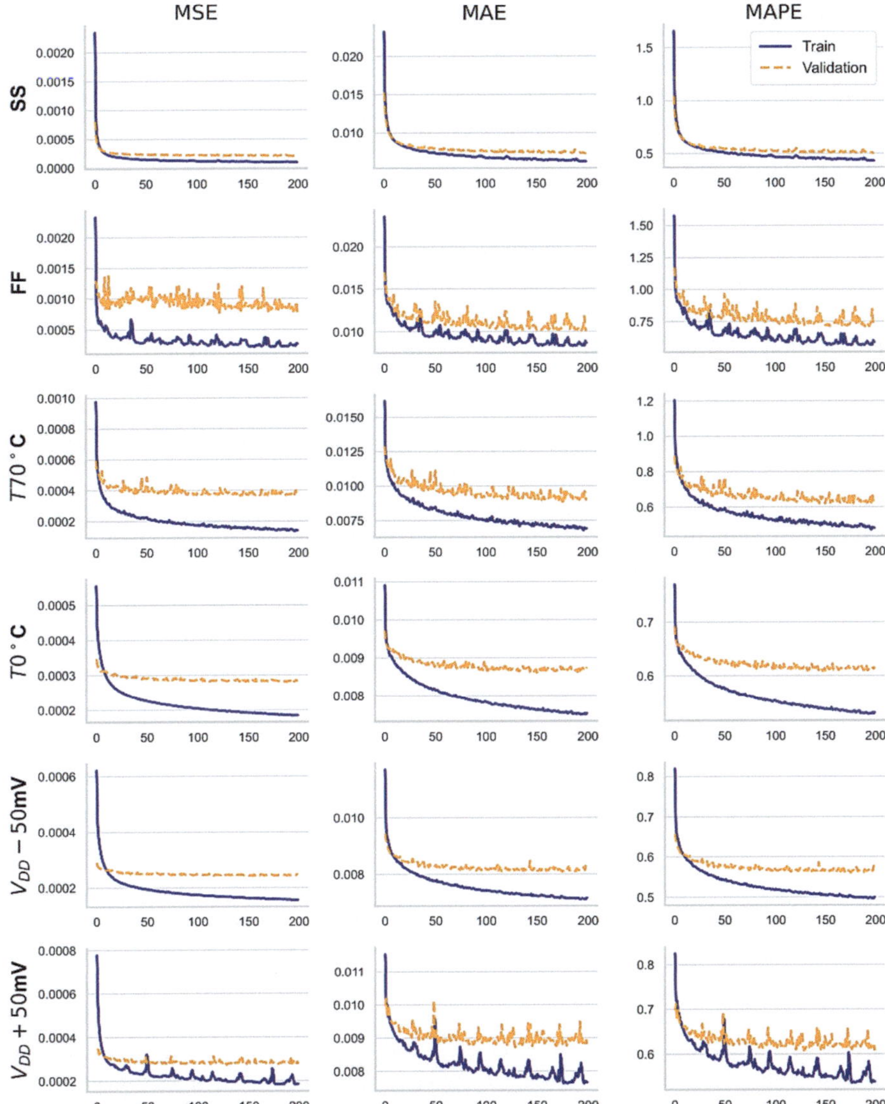

Fig. 4.20   Fine-tuning losses for the LNA3

optimization time from 130.12 h to just 46.30 h, translating into a speed-up factor of 2.81× over the original tool.

4.5 Conclusions and Future Research Directions

Recent advances in ML and DL techniques have been offering new alternatives to solve several design problems of analog ICs [27–31]. These include applications such as the inverse sizing problem, i.e., the mapping from specifications to the devices' sizes [32, 33], layout generation [34–37], fault testing [38], and so on. However, incorporating PVT corners into automatic analog IC synthesis is more critical than ever, and only a handful of works have reflected this growing importance [18, 39]. Therefore, in this Chapter, a novel automatic sizing optimization framework for analog and mmWave ICs was developed and thoroughly examined across several optimization runs on three distinct LNA topologies designed on a 65 nm CMOS node, on the challenging 28 GHz band. The proposed framework integrated a TL strategy from TT to PVT ANN-based performance regressors, and proved to be able to produce reliable PVT performance estimation models with approximately 83% less dataset generation effort than most recent efforts in the field [6]. The proposed PVTR framework was integrated into the simulation-based sizing tool of [20], and demonstrated speed-ups 2-to-3× over the former, while being able to produce solutions that meet all design requirements. The experiments conducted show that the specified TL strategy is well suited for mmWave design, with each transferred model being able to reach performance estimates with less than 10% deviation, setting an interesting baseline for future TL applications in analog and mmWave design.

4.5.1 Conclusions

In this Chapter, for the first time in the literature, the transfer of knowledge between TT and PVT schematic-level ANN-based performance regressors was tested and analyzed. This strategy aimed at reducing the dataset generation effort required to build accurate PVT performance estimators, a major bottleneck in IC design, due to the high simulation costs inherent in PVT corner-aware optimizations. Each transferred ANN was integrated into an existent simulation-based sizing tool, and two different methodologies were tested, where the models operated differently within the optimization loop. Table 4.15 summarizes the key conclusions from all experiments conducted, addressing the overall simulator workload reduction in each experiment performed, along with the speed-up granted by this methodology over the original automatic sizing tool. Overall, the results suggest that the developed TL-enhanced PVTR framework can be used to significantly accelerate analog and mmWave design, while also being able to produce sizing solutions that meet stringent performance requirements.

Table 4.15 Total speed-up achieved by the TL-enhanced PVTR framework

Circuit	PVTR	Base time [h]	Improved time [h]	Simulator workload reduction [%]	Speed-up [×]
LNA1	Semi-replace[1]	65.00	19.00	68.50	3.42
	Controlled[2]	128.71	55.14	52.80	2.33
LNA2	Semi-replace[4.1]	64.00	18.94	68.50	3.64
	Controlled[2]	127.33	61.65	34.51	2.06
LNA3	Controlled[2]	130.12	46.30	55.80	2.81

[1]Population size of 100 elements, optimized for 500 generations, with $\delta = 100$
[2]Population size of 100 elements, optimized for 1000 generations, with $\Delta = 10\%$ and $\delta = 100$

4.5.2 Future Work

As analyzed in Chap. 2, a variety of a DL techniques can be employed to accelerate the automatic sizing processes for analog and radio-frequency ICs. Among these approaches, ANN-based models have been extensively demonstrated as powerful tools for reducing the dependency on circuit simulators in the development of next generation EDA frameworks. Considering the methodology adopted in this Chapter, several perspectives for future research and possible methodological enhancements of the studied approach can be outlined as follow:

- Although the developed EDA framework was successfully validated using three distinct LNA circuits, further evaluation could be conducted on additional mmWave circuits, such as power amplifiers and voltage controlled oscillators. Such extension would further demonstrate the framework's versatility in handling different circuit classes and substantiate its plug-and-train functionalities;
- In recent years, fine-tuning methods, the process of retraining a base ANN using a set of pre-trained weights in a target domain, has become a mainstream technique within the transfer learning paradigm, and has been successfully applied across a wide range of applications [40]. Fine-tuning enables efficient knowledge transfer when source and target domains are closely related, while also providing strong initialization for adaptation in less similar scenarios. Consequently, evaluating the developed framework on different circuit classes would further demonstrate its robustness under various TT and PVT domain shifts. Moreover, domain adaptation techniques [41] could be explored to enhance generalization when the source and target domains exhibit significant dissimilarities;
- Several ML model families are appropriate for transfer learning, particularly those capable of hierarchical or shared representation learning. These obviously include feedforward neural networks, which enable efficient fine-tuning across related domains, as well as probabilistic models such as multi-task gaussian processes [42] that capture interdomain correlations. Even though two methodologies for adapting the pre-trained TT regressor were tested, additional experiments with

other techniques could be performed to discover the most well suited for these types of circuits.

References

1. Pan Z et al (2018) An efficient mm-wave integrated circuit synthesis method with accurate scalable passive component modeling. In: IEEE radio frequency integrated circuits symposium
2. Liu B et al (2012) An efficient high-frequency linear RF amplifier synthesis method based on evolutionary computation and machine learning techniques. IEEE Trans Comput Aided Des Integr Circ Syst 31(7):981–993
3. Passos F et al (2020) Synthesis of mm-wave wideband receivers in 28-nm CMOS Technology for automotive radar applications. IEEE Trans Comput Aided Des Integr Circ Syst 39(12):4375–4384
4. Yin S et al (2024) Automatic design for W-band front-end system via bottom-up sizing and layout generation. IEEE Trans Comput Aided Des Integr Circ Syst 43(3):705–715
5. Mendes L et al (2025) Fully automatically synthesized mm-wave low-noise amplifiers for 5G/6G applications. IEEE Trans Microw Theory Tech 73(8):4828–4841
6. Vaz P et al (2022) Speeding-up complex RF IC sizing optimizations with a process, voltage and temperature corner performance estimator based on ANNs. In: 2022 IEEE international symposium on circuits and systems (ISCAS)
7. Weiss K et al (2016) A survey of transfer learning. J Big Data 3(1)
8. Hongzhou L et al (2002) Remembrance of circuits past: macromodeling by data mining in large analog design spaces. In: Proceeding of DAC
9. Wolfe G, Vemuri R (2003) Extraction and use of neural network models in automated synthesis of operational amplifiers. IEEE TCAD 22(2):198–212
10. Alpaydin G, Balkir S, Dundar G (2003) An evolutionary approach to automatic synthesis of high-performance analog integrated circuits. IEEE TEC 7(3):240–252
11. İslamoğlu G et al (2019) Artificial neural network assisted analog IC sizing tool. In: Internal conference on SMACD
12. Çakıcı T et al (2020) Improving POF quality in multi objective optimization of analog ICs via deep learning. In: ECCTD
13. Taşkıran H et al (2023) ANN-powered reinforcement learning-based analog circuit optimization. IEEE ICECS
14. Oh Y et al (2024) CRONuS: circuit rapid optimization with neural simulator. In: Design, automation & test in Europe conference
15. Li Z et al (2024) An open-source AMS circuit optimization framework based on reinforcement learning—from specifications to layouts. IEEE Access 12
16. Wang Z et al (2022) Building a post-layout simulation performance model with global mapping model fusion technique. Tsinghua Sci Technol 27(3)
17. Wang Z et al (2022) Building post-layout performance model of analog/RF circuits by fine-tuning technique. In: Proceedings of the international symposium on quality electronic design (ISQED)
18. Mendes L et al (2021) In-depth design space exploration of 26.5-to-29.5-ghz 65-nm CMOS low-noise amplifiers for low-footprint-and-power 5g communications using one and-two-step design optimization. IEEE Access 9:70353–70368
19. Zhuang F, Qi Z et al (2020) A comprehensive survey on transfer learning. In: Proceedings of the IEEE, vol 109, no 1, pp 43–76
20. Martins R et al (2020) Design of a 4.2-to-5.1 GHz ultralow-power complementary class-B/C hybrid-mode VCO in 65-nm CMOS fully supported by EDA tools. IEEE Trans Circ Syst I: Regul Pap 67(11):3965–3977

21. Geron A (2022) Hands-on machine learning with Scikit-learn, Keras, and Tensorflow: concepts, tools, and techniques to build intelligent systems, 3rd edn
22. Hastie T, Tibshirani R, Friedman J (2009) The elements of statistical learning: data mining, inference, and prediction, 2nd edn
23. Kingma DP, Ba J (2015) Adam: a method for stochastic optimization. In: Proceedings of the 3rd international conference on learning representations (ICLR)
24. Chollet F et al (2015) "Keras". https://github.com/fchollet/keras
25. Abadi M et al (2015) TensorFlow: large-scale machine learning on heterogeneous systems. https://www.tensorflow.org
26. Chen C et al (2022) High-dimensional Bayesian optimization for analog integrated circuit sizing based on dropout and gm/ID methodology. IEEE Trans Comput Aided Des Integr Circ Syst 41(11)
27. Mina R, Jabbour C, Sakr G (2022) A review of machine learning techniques in analog integrated circuit design automation. Electron MDPI 11(3):435
28. Martins R, Lourenço N (2023) Analog integrated circuit routing techniques: an extensive review. IEEE Access 11:35965–35983
29. Wang C, Yang F, Zhu K (2024) AI-enabled layout automation for analog and RF IC: current status and future directions. In: Proceedings of the IEEE international symposium on radio-frequency integration technology
30. Maji S, Budak A, Poddar S, Pan D (2024) Toward end-to-end analog design automation with ML and data-driven approaches (invited paper). In: Proceedings of the 29th Asia and South Pacific design automation conference
31. Martins R (2025) A survey of machine and deep learning techniques in analog integrated circuit layout synthesis. Microelectron MDPI 1(1, 2)
32. Azevedo F, Lourenço N, Martins R (2025) Comprehensive application of denoising diffusion probabilistic models towards the automation of analog integrated circuit sizing. Expert Syst Appl 290:128414
33. Eid P, Azevedo F, Lourenço N, Martins R (2025) Using denoising diffusion probabilistic models to solve the inverse sizing problem of analog integrated circuits. AEU Int J Electron Commun 195:155767
34. Zhu K et al (2019) Genius route: a new analog routing paradigm using generative neural network guidance. In: Proceedings of the IEEE/ACM international conference on computer-aided design
35. Gusmão A, Horta N, Lourenço N, Martins R (2022) Scalable and order invariant analog integrated circuit placement with attention-based graph-to-sequence deep models. Expert Syst Appl 207:117954
36. Gusmão A, Póvoa R, Horta N, Lourenço N, Martins R (2022) DeepPlacer: a custom integrated OpAmp placement tool using deep models. Appl Soft Comput 115:108188
37. Gusmão A, Horta N, Lourenço N, Martins R (2021) Late breaking results: attention in Graph2Seq neural networks towards push-button analog IC placement. In: ACM/IEEE design automation conference
38. Andraud M, Stratigopoulos H, Simeu E (2016) One-shot non-intrusive calibration against process variations for analog/RF circuits. IEEE Trans Circ Syst I Regul Pap 63(11):2022–2035
39. Passos F et al (2018) Enhanced systematic design of a voltage controlled oscillator using a two-step optimization methodology. Integr VLSI 63:351–361
40. Guo Y et al (2019) SpotTune: transfer learning through adaptive finetuning. In: 2019 IEEE/CVF conference on computer vision and pattern recognition (CVPR), pp 4805–4814
41. Singhal P et al (2023) Domain adaptation: challenges, methods, datasets, and applications. IEEE Access 11:6973–7014
42. Liu B et al (2014) GASPAD: a general and efficient mm-wave integrated circuit synthesis method based on surrogate model assisted evolutionary algorithm. In: IEEE transactions on computer-aided design of integrated circuits and systems, vol 33, no 2, 169–181

 MIX
Papier aus verantwortungsvollen Quellen
Paper from responsible sources
FSC® C105338

If you have any concerns about our products,
you can contact us on
ProductSafety@springernature.com

In case Publisher is established outside the EU,
the EU authorized representative is:
**Springer Nature Customer Service Center GmbH
Europaplatz 3, 69115 Heidelberg, Germany**

Printed by Libri Plureos GmbH
in Hamburg, Germany